Organic Nitrogen in Drinking Water and Reclaimed Wastewater

Organic Nitrogen in Drinking Water and Reclaimed Waste Water

Prepared by:
Paul Westerhoff and Wontae Lee
Department of Civil and Environmental Engineering, Arizona State University
Tempe, AZ 85287-5306

Jean-Philippe Croue, Hervé Gallard
Laboratoire de Chimie de l'Eau et de l'Environnement
Universite de Poitiers, 86022 POITIERS Cedex - France

Gary L. Amy
Department of Civil, Environmental and Architectural Engineering
University of Colorado, Boulder, CO 80309

Sponsored by:
Awwa Research Foundation
6666 West Quincy Avenue
Denver, CO 80235-3098

Published by the
Awwa Research Foundation

DISCLAIMER

This study was funded by the Awwa Research Foundation (AwwaRF). AwwaRF assumes no responsibility for the content of the research study reported in this publication or for the opinions or statements of fact expressed in the report. The mention of trade names for commercial products does not represent or imply the approval or endorsement of AwwaRF. This report is presented solely for informational purposes.

CONTENTS

LIST OF TABLES

LIST OF FIGURES

FOREWORD

The Awwa Research Foundation is a nonprofit corporation that is dedicated to the implementation of a research effort to help utilities respond to regulatory requirements and traditional high-priority concerns of the industry. The research agenda is developed through a process of consultation with subscribers and drinking water professionals. Under the umbrella of a Strategic Research Plan, the Research Advisory Council prioritizes the suggested projects based upon current and future needs, applicability, and past work; the recommendations are forwarded to the Board of Trustees for final selection. The foundation also sponsors research projects through the unsolicited proposals process; the Collaborative Research, Research Applications, and Tailored Collaboration programs; and various joint research efforts with organizations such as the U.S. Environmental Protection Agency, the U.S. Bureau of Reclamation, and the Association of California Water Agencies.

This publication is a result of one of these studies, and it is hoped that its findings will be applied in communities throughout the world. The following report serves not only as a means of communicating the results of the water industry's centralized research program but also as a tool to enlist the further support of the nonmember utilities and individuals.

Projects are managed closely from their inception to the final report by the foundation's staff and large cadre of volunteers who willingly contribute their time and expertise. The foundation serves a planning and management function and awards contracts to other institutions such as water utilities, universities, and engineering firms. The funding for this research effort comes primarily from the Subscription Program, through which water utilities subscribe to the research program and make an annual payment proportionate to the volume of water they deliver and consultants and manufacturers subscribe based on their annual billings. The program offers a cost-effective and fair method for funding research in the public interest.

A broad spectrum of water supply issues is addressed by the foundation's research agenda: resources, treatment and operations, distribution and storage, water quality and analysis, toxicology economics, and management. The ultimate purpose of the coordinated effort is to assist water suppliers to provide the highest possible quality of water economically and reliably. The true benefits are realized when the results are implemented at the utility level. The foundation's trustees are pleased to offer this publication as a contribution toward that end:

Walter J. Bishop
Chair, Board of Trustees
Awwa Research Foundation

Robert C. Renner, P.E.
Executive Director
Awwa Research Foundation

ACKNOWLEDGMENTS

The authors of this report are indebted to the following water utilities and other partners for their cooperation and participation in this project:

- Shimadzu Scientific (Columbia, MD)
- City of Phoenix (Phoenix, AZ)
- Louisville Water Company (KY)
- Southern Nevada Water Company (Las Vegas, NV)
- US Filter / Vivendi (Atlanta, GA)
- Contra Costa Water District
- Milwaukee Water Works
- City of Ann Arbor (MI)
- City of Fort Worth (TX)
- City of Philadelphia (PA)
- City of Gilbert (AZ)
- Central Lake County Joint Action Water Authority
- City of Peoria (AZ)
- Massachusetts Water Resources Authority (Southborough, MA)
- Joplin WTP (MO)
- Jefferson City WTP (MO)
- St. Louis County WTP (MO)
- Newport News Waterworks (Yorktown, VA)
- American Water (Vorhees, NJ) participating facilities: Alton WTP (IL), East St. Louis WTP (IL), Granite City WTP (IL), Peoria District WTP (IL), Streator WTP (IL), Pontiac WTP (IL), Kokomo (IN), Richmond (IN), Muncie (IN)

In addition, Peter Fox (Professor), Heath Mash (Research Associate), Mario Esparza-Soto (Research Associate), and Robert Marzke (Professor) from Arizona State University provided invaluable technical input and direct research assistance on this project. As graduate students at Arizona State University, Baiyang Chen and Joby Carlson contributed to the success of this project through laboratory work and discussions. Their advice and help, plus the advice of the Project Advisory Committee (PAC) – including Tanju Karanfil (Clemson University), Jonathan Sharp (University of Delaware), Laxman Davkota (City of Phoenix) – and the help of the AwwaRF project officer, Djanette Khiari, are appreciated. Advice from Jerry Leenheer (USGS, Denver, CO) is also very useful.

EXECUTIVE SUMMARY

RESEARCH OBJECTIVES

The goal of this first AwwaRF project on dissolved organic nitrogen (DON) is to document DON occurrence and provide information on its chemical characteristics and reactivity towards metal hydroxides and oxidants/disinfectants. In the past, relatively minor attention has been focused on DON despite observations that both low and high molecular weight molecules containing organic nitrogen (e.g., simple amino acids, peptides, algal-derived humic substances) have been implicated as precursors of disinfection by-product (DBP) such as trihalomethanes (THMs), haloacetic acids (HAAs), haloacetonitriles (HANs), nitromethanes, and nitrosamines like nitrosodimethylamine (NDMA). Nitrogenous organic matter has also been implicated as membrane foulant materials. Dissolved organic carbon (DOC) is usually used as the surrogate for natural organic matter (NOM); NOM contains various ratios of carbon, oxygen, hydrogen, nitrogen, sulfur and trace metals. While information exists on NOM removal through DOC measurements, it is unclear if DON is distributed equally among different molecular weight fractions. For example, while higher organic nitrogen content is associated with increasingly polar natural organic matter (NOM) fractions, there are no publications of DON removal during coagulation. The research uses both well-developed DON analytical methods from the fields of ecology and oceanography and innovative DON quantification methods tailored for the drinking and reclaimed wastewater industry to meet the following specific objectives:

- Develop an accurate DON quantification method suitable for drinking waters
- Develop a comprehensive assessment of DON occurrence in raw/finished drinking waters and reclaimed wastewaters
- Evaluate a suite of analytical tools for characterizing DON structure
- Quantify DON removal by sorption processes (coagulation, activated carbon)
- Investigate role of DON on DBP formation (HAN, NDMA, THM) during disinfection
- Validate methods for organic chloramine measurement and assess organic chloramine formation during disinfection (free chlorine and monochloramine)
- Assess future organic-nitrogen research needs.

APPROACH

After developing and verifying methods for accurate determination of DON in drinking water and reclaimed wastewater, a dialysis-based pretreatment method was applied for assessing DON occurrence at water treatment plants (WTPs) and surface waters across the United States and France. Techniques to characterize nitrogenous components of NOM were developed and applied to some of the field samples or to NOM fractions isolated during previous AwwaRF projects. Raw waters and NOM isolates were used in bench-scale coagulation, adsorption, and oxidation (free chlorine and monochloramine) studies to assess DON removal and reactivity in forming organic chloramine and DBPs.

CONCLUSIONS

This study developed a dialysis-based pretreatment method and validated a high-temperature oxidation analytical method that is suitable for measuring DON in drinking waters. A survey of 28 full-scale water treatment plants found an average DON concentration of 0.19

mgN/L in raw waters and 0.15 mgN/L in finished water, and DOC/DON ratios of raw water averaged 18 mgC/mgN. DON removal during water treatment closely mirrored DOC removal patterns. DON molecular weight distributions are similar to those for DOC. Coagulation preferentially removed the higher molecular weight DOC and DON materials. Addition of cationic polymers improved DON removal at low to moderate alum dosages. Complex (hydrolysable) amino acids and aminosugars can account for the majority of structural groups that contain organic nitrogen. Polar and colloidal NOM fractions are the most nitrogen enriched. Free chlorine and monochloramine react with nitrogenous NOM and form organic chloramine and DBPs. Organic chloramine may interfere with disinfectant measurements and serve as DBP precursors. A source for nitrogen in nitrogenous DBPs can be DON present in NOM. Specific findings from the study are highlighted below.

DON Measurement

Prior to this project methods for DON determination suffered from high reporting levels that were not suitable for application to drinking water studies and/or were inaccurate in the presence of elevated inorganic nitrogen species such as nitrate or ammonia. A direct DON measurement method would remove dissolved inorganic nitrogen (DIN) species with minimal loss of organic nitrogen, or an improved DON measurement method would increase the accuracy of DON by sufficiently lowering the DIN concentration. In drinking waters nitrate is the dominant DIN species Three approaches were evaluated to improve the accuracy of DON determination, which would be suitable for drinking water and reclaimed wastewater:

1. Catalytic reduction of nitrate and nitrite to nitrogen gas or ammonia
2. Adsorption of ammonium ion onto zeolite (e.g., after catalytic reduction of nitrate)
3. Permeation of nitrate, nitrite, and ammonia across dialysis membranes

Catalytic Nitrate Reduction

- Several Pd-Sn/Al_2O_3 catalysts were evaluated that were previously shown to reduce nitrate
- Catalytic reduction of two free amino acids (methionine and cystine) occurs, producing alanine as by-products. Most free amino acids did not react significantly with the catalysts, although for most amino acids a 5% to 15% lower concentration was measured after contact with the catalysts (possibly due to adsorption).
- Initial tests using catalysts produced from ammonium nitrate salts indicated low amounts of residual inorganic nitrogen, so the final set of catalysts were prepared instead from chloride salts.
- In NOM-free water the catalysts rapidly reduced nitrate (0.6 mM/g-min) to volatile nitrogen gases (81%) and ammonia (19%) at high nitrate concentrations. The selectivity of the catalysts to produce nitrogen gases decreased as initial nitrate concentrations were lowered; this unfortunately increased the yield of ammonia.
- In the presence of NOM isolates an approximate 10% to 20% decrease in DON was observed, due to adsorption and/or reduction.
- In the presence of NOM isolates, nitrate reduction proceeded more slowly than in NOM-free water. Additionally, nitrate was stoichiometricly reduced to ammonia without formation of volatile nitrogen gases.

- Overall catalytic processes were not considered feasible for improving the accuracy of DON determination in drinking waters.

Ammonium Ion Adsorption onto Zeolites

- Zeolites removed >90% of the initial ammonia added to distilled water or tap water. Volumetric zeolite dosages of 2% to 10% were required to achieve these removal levels. Lower zeolite dosages had < 90% ammonia removal.
- Equilibrium conditions for ammonium ion removal by zeolites were achieved with 60 minutes of contact time.
- At zeolite dosages required for high ammonia removal (2% to 5% volumetric dosages), approximately 5% to 10% of the DOC and DON were adsorbed.
- Zeolites can be used to remove ammonia from high ammonia waters (e.g., nitrified wastewaters), but are not suitable for improving the accuracy of DON determination in drinking waters, without first applying catalytic reduction for nitrate, since they usually contain low ammonia levels.

Dialysis Pretreatment for DIN/DON Separation

- Dialysis membranes with a nominal molecular weight cutoff of 100 dalton allowed passage of nitrate, nitrite, and ammonia across the membrane.
- A few low molecular weight organic nitrogen compounds (e.g., urea with a MW of 96 g/mole) passed across the dialysis membrane, but compounds with MW>120 g/mole did not pass across the membrane.
- A potential exists for hydrophobic organic nitrogen compounds or colloidal NOM isolates to adsorb onto the dialysis membrane material, but under the conditions studied (2 cm^2 membrane per mL of sample) this adsorption accounted for <10% of the DOC or DON.
- Dialysis against a continuous flow of ultrapure water resulted in near complete elimination of nitrate, nitrite, or ammonia while retaining >95% of the DOC.
- Depending upon dialysis duration, DIN will be completely or partially removed. A dialysis duration of 24 hours was selected to maximize DIN permeation while minimizing DOC loss to <5%.
- In many cases DIN was removed to below detection limits, allow for direct quantification of DON through TDN techniques. However, if initial nitrate, ammonia, or nitrite are above 1 mgN/L then their concentrations should be determined post-dialysis.
- A dialysis pretreatment protocol was developed for use with drinking waters or reclaimed wastewater with DOC concentrations less than 6 mg/L.
- Below a DIN/TDN ratio of 0.6 mgN/mgN, DON accounts for >40% of the measured TDN and accurate DON determination can be achieved without dialysis pretreatment, as long as the TDN and DIN specie analytical methods have appropriate method detection levels (MDLs).
- Above the DIN/TDN ratio of 0.6 mgN/mgN, cumulative analytical errors in nitrate, nitrite, and ammonia measurements decrease the accuracy of DON determination. Dialysis pretreatment was found to statistically improve DON accuracy for waters with DIN/TDN ratios above 0.6 mgN/mgN.

DON Occurrence

Dialysis pretreatment with a high temperature catalytic oxidation technique for nitrogen analysis (Shimadzu TOC-V) was used for water treatment plant sampling since the average DIN/TDN was greater than 0.6 mgN/mgN. The following observations were made from a study of two sampling campaigns conducted at 28 WTPs in the US:

- The mean DON concentration for 28 water treatment plants in the US was 0.19 mgN/L for raw waters and 0.15 mgN/L for finished waters. The DON concentrations ranged from <0.05 to 0.45 mgN/L.
- The mean concentrations were below the 0.2 mgN/L MDL for Total Kjeldahl nitrogen (TKN) measurements. This suggests that other methods, such as the high temperature catalytic oxidation technique used herein, with lower MDLs are required for wide-spread use in the drinking water industry.
- The average DOC/DON ratio for the raw waters was 18mgC/mgN.
- Average DON removal across the 28 WTPs was 20%, which was lower than the average DOC removal (29%) or UVA_{254} removal (55%). WTPs using polymers had approximately a 9% higher DON removal than WTPs not reporting use of polymer during coagulation.
- Seasonally DON concentrations increased in the epilimnion of water supply reservoirs, during warmer water temperatures and periods of algal activity that produced soluble microbial products. In some reservoirs the DIN/TDN ratio effectively decreased to zero, as all the dissolved nitrogen became incorporated as DON.
- Within common watersheds or similar hydrologic settings, higher DON concentrations correlated with increasing water temperatures. This suggests temperature may be an indicator for DON variability. This observation could result from algal and bacterial activity, or changes in hydrologic flowpaths.

For two WTPs in the US and France, most of the DON removal occurred during coagulation. For WTPs with granular activated carbon, additional removal was observed. For a WTP in France with coagulation, filtration, and ozonation the DON removal was 25% and comparable with DON removals observed at the US WTPs. For a second WTP in France with additional treatment processes (flotation, GAC), higher DON removals (53%) were observed. At a few US WTPs that practiced chloramination prior to filtration (e.g., excess free ammonia present), a transformation of free ammonia to DON during filtration was observed and attributed to bacterial processes.

In both the US and France discharging wastewater into surface waters (reservoirs or rivers) increases DON concentrations. Wastewater contains higher DON than non-impacted surface waters. Furthermore the addition of nutrients in wastewater effluents increases the potential for algae production, and subsequent formation of soluble microbial products that contain DON. Discharge of wastewater into the subsurface (i.e., soil aquifer treatment) decreases the DON and DOC concentrations. The magnitude of these reductions depends upon travel time.

DON Characterization

The molecular weight and chemical composition of DON in drinking waters and reclaimed wastewaters was poorly documented. The following conclusions regarding characterization of nitrogenous organic materials in drinking waters were reached:

- The carbon to nitrogen ratio for NOM in drinking water ranges between 10 to 30 mgC/mgN. Transphilic and colloidal NOM isolate fractions are more nitrogen enriched (DOC/DON between 5 to 10 mgC/mgN) while hydrophobic acids generally have higher DOC/DON ratios.
- Based upon dialysis, >95% of the DON has molecular weights greater than 100 dalton.
- Based upon ultrafiltration separation, the molecular weight distribution of DOC and DON in drinking waters are comparable. In raw drinking water, DON was equally distributed (~ 1/3rd each) into the three ranges considered: < 1000 dalton, 1000 to 10,000 dalton, and >10,000 dalton. In finished drinking water, the distribution was skewed towards the lower molecular weight fractions, with approximately half of the DON < 1000 dalton and then equal amounts of the other two molecular weight ranges.
- High performance size exclusion chromatography (HPSEC) with DOC, UV absorbance at 254 nm, and fluorescence on-line detection indicated that colloidal NOM isolates contain high molecular weight biopolymers. The data also confirms that most of the DOC has molecular weights greater than that of free amino acids. Analysis of the HPSEC with fluorescence detection data suggests the fluorescence is attributed to aromatic materials other than tryptophan or similar low molecular weight nitrogenous aromatic molecules.
- HPLC analysis with ultraviolet absorbance (UVA) detection at multiple wavelengths (210nm to 254 nm ratio of absorbances) was used to compute the ultraviolet absorbance ratio (URI) value of NOM. The URI value is indicative of proteinacous materials. Two HPLC peaks corresponding with protein/polysaccharides and low molecular weight materials exhibited high URI values, and considered to be representative of nitrogenous materials.
- Simple spectroscopy measurements (UVA, fluorescence intensity), with HPLC, also provided reasonable correlations to DON or DOC/DON ratios. After correcting UVA_{210} for the presence of nitrate, the ratio of corrected UVA_{210} to UVA_{254} correlated with DON. Fluorescence intensity at wavelengths indicative of protein-like fluorophores (excitation 270-280 nm and emission 300-350 nm) correlated with DON concentrations.
- Methods were applied for amino acid and aminosugar quantification. Using methanesulfonic acid for hydrolysis of NOM isolates yielded higher amino acid content compared to the use of hydrochloric acid for hydrolysis. For a colloidal NOM isolate methanesulfonic acid (MSA) hydrolysis yielded an amino acid concentration equivalent to 584 μgN/L, or 53% of the DON, compared to only 316 μgN/L, or 30% of the DON, by HCl hydrolysis. Similar improvement in hydrolysis yields were observed for a TPI NOM isolate, where MSA yielded amino acids that accounted for 21% of the DON compared to only 8% after HCl hydrolysis.
- Different NOM isolates exhibited variable levels of amino acids. Glutamic acid, glycine, lysine and arginine were among the highest concentration in many samples. At the time only HCl hydrolysis was being employed. The colloidal fraction was most enriched with aminosugars. The HPO isolate was more enriched in acidic and aromatic amino acids

compared against a TPI fraction from the same water source. The TPI fraction was enriched in neutral amino acids and aminosugars.

- Pyrolysis GC-MS analysis requires desalting of bulk waters, which can result in significant loss of NOM. Future work should consider the use of dialysis as a desalting step rather than resin extraction. The protein, peptides, and aminosugar fractions determined by pyrolysis GC-MS accounted for approximately 3% to 5% of the signal. This is in reasonable agreement with nitrogen content of NOM.
- ^{15}N-NMR was investigated as a potential means to augment standard FTIR analysis. However, due to the low nitrogen isotopic content of NOM very long NMR run times are required to acquire meaningful signals, although it was quite easy for model nitrogen compounds. After acquiring ^{15}N-NMR spectra for a IHSS soil humic acid spectra peaks indicative of amine bonds were observed. Overall, ^{15}N-NMR may hold promise for nitrogenous NOM characterization but for the amount of effort required to obtain sufficient NOM mass for analysis, FTIR analysis could yield comparable information with less than 1/10th the mass of material.
- Several commercially available protein test-kits were evaluated. The Lowry and Bradford assays are appropriate for untreated or poorly treated wastewaters, but are not sensitive enough for low protein concentrations such as found in reclaimed wastewaters or drinking waters. A biomedical assay (NanoOrange) used to measure protein content in cells is very sensitive, but exhibited interferences from salts in water, especially from divalent ions. As such it is of limited use for bulk drinking water samples, but could be used on samples after dialysis pretreatment. For NOM isolates, the NanoOrange method could be used without concern for salts. The colloidal fraction exhibited the highest protein content based upon the NanoOrange reagent.

DON Removal in Bench-Scale Tests

To aid in understanding DON removal patterns observed in full-scale WTPs during the DON occurrence sampling campaigns, bench-scale tests were conducted. Coagulation and activated carbon adsorption tests were conducted. The following trends were observed:

- Coagulation with aluminum salts removes both DOC and DON. Depending upon the source water and aluminum dosage, DON removal can be equal to, greater than, or less than DOC removal upon coagulation. For waters with higher specific ultraviolet absorbance at 254 nm (SUVA), DON removal is equal to or even higher than DOC removal. Reclaimed wastewaters had low SUVA values (< 2 L/mg-m), and resulted in low overall DON removal and less DON than DOC removal.
- Cationic polymers are frequently used in conjunction with metal salt coagulants. PolyDADMAC polymers contain 7% organic nitrogen, so their use actually adds DON at the beginning of the coagulation process. However, the use of low polyDADMAC dosages improved DON removal by 15% to 20% during coagulation at commonly used alum dosages, compared against the absence of the cationic polymer. It is hypothesized that the cationic polymer neutralizes anionic charge on more polar NOM fractions, which tend to be nitrogen enriched over the hydrophobic NOM fractions.
- Coagulation preferentially removed higher molecular weight fractions of DON and DOC. These higher molecular weight fractions have higher SUVA values than the lower molecular weight fractions. Coagulation appears to preferentially removes higher

molecule weight, higher SUVA organic matter that contains both carbonaceous and nitrogenous materials.

- DON can adsorb onto activated carbon. While reclaimed wastewater nitrogenous materials are difficult to coagulate, they exhibit a higher tendency to adsorb onto activated carbon than DON from a surface water.

Organic Chloramine Formation

Organic chloramine can interfere with detection of inorganic chloramine detection and may serve as intermediates for THM, HAN, or NDMA. The kinetics and yields of organic chloramine were assessed as a first step in understanding the potential presence of organic chloramine in drinking water.

- Organic chloramine are detected by the Total DPD colorimetric method. The indophenol method (Hach MonoChlorF) does not react with organic chloramine, and appears much better for monitoring inorganic chloramine residuals than Total minus Free DPD colorimetric methods, for utilities practicing chloramination.
- Organic chloramine form in the presence of chlorine. The kinetics of organic chloramine formation includes a rapid reaction phase (< 5 minutes) and a slower formation phase lasting up to approximately 24 hours.
- The yield of organic chloramine increased linearly as a function of chlorine consumed (slope ~ 0.05 mg Cl_2/L of organic chloramine produced per unit free chlorine consumed (mg Cl_2/ mg DON). For example, a water with 0.2 mgN/L with a 2 mg Cl_2/L free chlorine demand could form 0.5 mg Cl_2/L of organic chloramine.
- Organic chloramine form in the presence of monochloramine, although the reaction rates are slower than for free chlorine. The maximum yield of organic chloramine formation upon addition of monochloramine is only 1/3rd the yield upon free chlorine addition.
- Organic chloramine form when exposed to free chlorine for a short time (20 to 120 minutes) prior to ammonia addition. This is a common scenario for full-scale water utilities using chloramine as a post-disinfectant.
- Different NOM isolates had variable DOC/DON ratios, but yielded comparable levels of organic chloramine when normalized to DON content of the isolates. Slight differences in organic chloramine yield could be attributed to variable chlorine demand by the isolates.
- After organic chloramine formation, hydrolysis appears to be responsible for its slow decay over time. Half-lives for organic chloramine were on the order of 12 to 24 hours.

Regulated and Nitrogenous DBP Formation

Limited experimentation conducted on DBP formation focused at contrasting HAN and NDMA formation during chlorination and chloramination of 18 NOM isolates with a range of DOC/DON ratios. No bromide was present. THMs were also measured. The following observations were made:

- NDMA did not form during chlorination of nitrogen containing NOM isolates.
- NDMA did form during chloramination of NOM isolates. This observation confirms that NOM can serve as precursor material for NDMA under drinking water conditions. The

colloidal NOM isolate, and other nitrogen enriched isolates, had the highest NDMA yield compared against hydrophobic acid isolates.

- The ratio of THM/HAN increased as a function of the DOC/DON ratio. Therefore, NOM isolates with higher nitrogen content had lower THM and higher HAN formation compared against NOM isolates with less nitrogen content.

RECOMMENDATIONS FOR FUTURE RESEARCH

This study allows utilities to measure DON and understand its role during treatment. The research will allow utilities to understand immediate and longer-lasting issues, including the following:

- **DON measurement**. In most cases TKN analysis will not provide low enough detection limits for monitoring the change in DON concentration in raw or finished waters. Utilities should monitor DON seasonally to begin collecting data that might influence disinfection practices (e.g., conversion to chloramine) or formation of emerging nitrogenous DBPs (haloacetonitriles, nitromethanes, nitrosamines).
- **Expected DON concentrations**. The average raw water DON concentration observed in this study was 0.19 mgN/L and was close to the average value of 0.3 mgN/L reported previously by the Investigators after analysis of USGS surface water data. Groundwaters usually have lower concentrations. Upstream agricultural activities, wastewater discharges, or significant algae productivity will result in higher DON concentrations. The average DOC/DON ratio was 18 mgDOC/mg DON and could be used as a rough estimate of the DON concentration of raw water samples.
- **Surrogate measurements for DON**. In addition to DOC measurements, other surrogate measurements for DON concentrations were explored. Lower SUVA tends to correlate with NOM that is enriched with organic nitrogen compared to higher NOM with a higher SUVA. Fluorescence spectroscopy appears to be a reasonable surrogate for DON, and reasonably priced instrumentation can be purchased. Specific excitation and emission wavelengths tend to correlate with protein-like material. Other surrogates (Lowry Method, Bradford Method, NanoOrange method) have inadequate MDLs or potential inorganic interferences.
- **Composition of DON**. Large macromolecular compounds contain amino acid and aminosugar components, which account for up to 60% of the DON. The concentration of free amino acids are less than 5% of the DON. NOM fractions have different nitrogen content. The colloidal NOM fraction generally has the highest nitrogen content, and is indicative of biological activity upstream. More polar (less hydrophobic) fractions are also nitrogen enriched, so humic substances tend to have less DON than NOM not retained on XAD8 resins.
- **Removal of DON**. Coagulation with metal salts have equal or slightly lower DON percentage removals compared to DOC removals. High molecular weight DON, and DOC, is preferentially removed during coagulation. To improve DON removal at low alum dosages, cationic polymers can be used. DON of biological origin (e.g., wastewater discharges) tends to be difficult to coagulate, but can exhibit a moderate tendency to adsorb to activated carbon (e.g., GAC). Membrane tests were not specifically conducted during this study, but based upon the ultrafiltration molecular weight fraction conducted,

the distribution of DON and DOC among molecular weight fractions were quite similar. Therefore, membrane processes capable of removing DOC should also remove DON.

- **Chlorination or Chloramination of DON**. Organic chloramine form due to reactions between DON and these oxidants. Organic chloramine interfere with Total DPD colorimetric measurements of combined chlorine residuals. The indophenol method (Hach MonoChlorF reagent) did not exhibit interference from organic chloramine, and should be used for monitoring. Utilities may want to recommend to state to adopt the indophenol method for monitoring monochloramine. Organic chloramine formation in the presence of free chlorine occurs rapidly within a few minutes, whereas formation in the presence of preformed monochloramine is much slower and did not reach a maximum concentration until nearly 24 hours after exposure to the monochloramine. This is significant, because utilities using short free chlorine contact times for in-plant microbial inactivation will form organic chloramine, before ammonia addition to form monochloramine as a post disinfectant. Once formed organic chloramine degrade slowly over days due to hydrolysis.
- **Organic Chloramine Formation.** What are organic chloramine concentrations in water distribution systems, and does organic chloramine formation lead to negative water quality impacts? The Investigators can hypothesize a sequence of events leading to nitrification in water distributions for utilities practicing chloramination. An algae bloom occurs in the spring or late summer in a watershed. The DOC/DON ratio decreases, but the absolute DOC concentration may only change slightly. DON concentrations increase in treated water. Free chlorine is added for disinfection, forming organic chloramine. Then free ammonia is added to form inorganic chloramine for post disinfection, but organic chloramine are already present and continue to form slowly due to reactions between DON and monochloramine. With travel time away from the WTP total DPD measurements indicate a 1 mgCl_2/L combined residual, but > 75% of that residual could be due to organic chloramine or dichloramine. In the presence of these weak biocides, nitrifying bacteria proliferate. The remaining monochloramine reacts with the biofilm (nitrogen rich) and may form additional organic chloramine, that remain detected by total DPD detection. Then nitrification occurs and all residual is lost. The common remedy then includes several weeks of free chlorination. Do organic chloramine form during this period as free chlorine reacts with the biofilms? Does NDMA or other nitrogenous form? Much of this potential problem could have been avoided if utilities utilize and state regulators adopt the indophenol method for monochloramine detection and monitoring in water distribution systems.
- **Nitrogenous DBPs**. Free chlorination of NOM formed HANs and THMs, whereas chloramination formed higher HAN but lower THM concentrations. In the presence of free chlorine, HANs can serve as intermediates and ultimately form THMs. Free chlorination of NOM isolates did *not* form NDMA, but NDMA did form in the presence of monochloramine. Overall DON is a precursor for nitrogenous DBPs (HAN, NDMA) as well as currently regulated DBPs (THMs). Through monitoring DON, a better appreciation for the presence of nitrogenous DBP precursors will be gained.

CHAPTER 1 INTRODUCTION AND OBJECTIVES

STATEMENT OF PROBLEM

Dissolved organic nitrogen (DON) has become an issue to the water industry (drinking water and reclaimed wastewater) primarily because of its potential to react with oxidants/ disinfectants and form disinfection by-products of health concern (e.g., nitrosamines, nitromethanes, acetonitriles, trihalomethanes, etc), and its potential role in membrane fouling. The Investigators recently published a review article on organic nitrogen that discusses DON occurrence and characterization of DON, methods for DON analysis, and the treatability of DON as related specifically to drinking water treatment (Westerhoff and Mash 2002). DON constitutes a relatively small fraction of the dissolved organic material, with carbon to nitrogen ratios (C/N or DOC/DON) ranging between 5 to 100 mgC/mgN. DON concentrations range from < 0.1 to > 5 mgN/L. Surface waters generally have higher DON concentrations than groundwaters, unless the groundwater has high DOM levels or the groundwater is located beneath specific land use types (e.g., wastewater recharge, agricultural fertilization). DON concentrations generally follow similar seasonal patterns as dissolved organic carbon (DOC), are influenced by hydrologic flowpaths and algae, wastewater or manure sources. As such the DOC/DON ratios may be good surrogates for determining the origin of DOM in watersheds. Amino acids constitute roughly 15%-70% of the DON, while the remainder consists largely of uncharacterized amide, nitro, nitrile and heterocyclic nitrogen. DON removal and effects on treatment performance during drinking water treatment (oxidation, metal coagulation, membrane separation) have not been widely reported and are generally ill-defined. Research is needed that characterizes the non-amino acid nitrogenous compounds, characterizes bulk DON properties (e.g., molecular weight), and identify surrogates (e.g., fluorescence) for DON occurrence and fate and DON removal during water treatment.

Organic nitrogen is commonly the dominant form (>50%) of dissolved nitrogen in water, but comprises a relatively small portion (0.5% to 10% by weight) of the mass of natural organic matter (NOM) or dissolved organic matter (DOM). Dissolved organic carbon (DOC) has traditionally been used as a surrogate for NOM; NOM contains roughly 30% to 50% carbon by weight. Conceptual models for NOM structures include aromatic and aliphatic carbon with carboxyl, hydroxyl or carbonyl function groups, but often lack significant inclusion of org-N structures. While exclusion of org-N structures may be associated with the generally low elemental N composition of NOM, other than amine functionality very little is known about org-N structures in NOM (e.g., pyroles, amides, nitriles, purines, pyrimidines, nitro compounds). Despite considerable information on various aspects of DOC cycling in the environment or fate during water treatment, little is known about the nitrogenous fraction of NOM. Org-N exerts oxidant / disinfectant demands (e.g., chlorine, ozone) and can play the role of disinfection by-product (DBP) precursors. Removal of particulate organic nitrogen (PON) poses treatment challenges (e.g., removal of algal cells), as does dissolved organic nitrogen (DON). During water treatment, biological treatment processes can produce DON (e.g., soluble microbial products (SMPs)), and DON can affect the performance of other physico-chemical processes (e.g., contribution to membrane fouling). Very few direct measurements of DON are available from drinking water systems. Kjeldahl nitrogen (TKN) measurements (org-N + NH_3) in wastewater are not sensitive at < 1 mgN/L.

An abundance of information has been published on the occurrence of organic nitrogen in marine and freshwater ecosystems. However, a detailed study of the occurrence and treatability of DON during drinking water treatment is not available. Planning workshops organized over the past decade by Awwa Research Founding all identified org-N as a potentially critical issue for drinking water treatment (AwwaRF 1993; AwwaRF 2001). The European Community directive has a maximum admissible concentration for organic nitrogen of 1 mg Kjeldahl-N/L, but the USA has no similar guideline related to org-N. Emerging concerns over increased contribution of potentially elevated DON levels from upstream wastewater or agricultural sources in drinking water supplies and identification of potent carcinogenic nitrogenous disinfection by-products (e.g., nitrosamines, nitromethanes) increases the need to better understand the concentrations, structures and reactivity of nitrogenous organic components in drinking water systems.

Organic Nitrogen Analysis

Total organic nitrogen (TON) is measured after a preparative step in which all the organic material is converted (combusted or oxidized) to an inorganic nitrogen species (NO_2^-, NO_3^- or NH_3/NH_4^+) for measurement in solution or into gaseous nitrogen oxides (NO_x). Inorganic nitrogen species are typically present in natural waters prior to this step (NH_4^+, NO_2^- and/or NO_3^-), which necessitate measuring the total concentration of inorganic nitrogen specie(s) prior to and following the conversion step(s). Therefore, organic nitrogen is commonly quantified by the difference between total nitrogen (TN) and dissolved inorganic nitrogen (DIN), which is the sum of NH_4^+, NO_2^- and NO_3^-, as follows:

$$TON = TN - DIN \quad (1.1)$$

$$DON = TDN - DIN \quad (1.2)$$

where TN is measured on unfiltered samples and total dissolved nitrogen (TDN) is analyzed on filtered aliquot { Overall classification: TON = PON + DON; TIN $\approx$ DIN}. Measurement of org-N commonly involves three steps:

1. **Measurement of inorganic nitrogen species** – The methods must have low analytical error since inorganic nitrogen is subtracted from total nitrogen (Equations 1.1 & 1.2) to determine org-N levels. Nitrate, nitrite, and ammonia analytical methods include ion selective electrodes, ion chromatography, vanadium reduction, spongy cadmium method, cadmium reduction, hydrazine reduction, Devarda's alloy reduction, titanous chloride method, UV spectroscopy, and colorimetric determinations.
2. **Conversion of org-N to inorganic nitrogen** - Three broad methods have been developed: wet-oxidation, high temperature oxidation and photolytic oxidation. Both wet and photolytic oxidations rely on subsequent measurement of inorganic nitrogen species in solution, while high temperature oxidation methods measure nitrogen oxide indirectly through ozonation and chemi-luminescence detection. Comparison of the methods for model organic compounds yielded nitrogen recoveries of >90% to 110%. Extracted humic acids and natural water samples also had similar results among the methods, although photolytic oxidation methods show higher levels of variability with possible lower recoveries compared to the other methods.
3. **Second measurement of inorganic nitrogen**. Same as #1 above, but now includes org-N material oxidized to inorganic material (#2 above). Calculation of DON by Equation 1.2.

Traditionally wet chemical methods have been widely applied in the fields of ecology and marine sciences, probably offering sufficient resolution for the drinking water industry. Several partially or fully automated digestion of inorganic and organic species and colorimetric in-line analyzers are commercially available for these measurements (e.g., Bran & Luebbe TrAAcs 800 Autoanalyzer, Lachat Quick Chem 8000 Flow Injection Analyzer). These systems have method detection limits (MDLs) for the inorganic ions of the order of 5 μg/L, but due to the digestion method and DON computation by difference (Equation 1.2), the MDLs for DON is higher (~ 20 μg/L). Many drinking and reclaimed wastewaters can have nitrate levels of the order of 3 to 8 mgN/L, necessitating DON computation (a small number) by subtraction of two large values (TDN and DIN). New high temperature digestion systems with on-line detection are also commercially available (Shimadzu, Antek, Dohrmann, Euroglas), but such systems only quantify TN or TDN (filtered samples) and not DON directly. However, if the addition of an efficient DIN removal pretreatment system was developed prior to high-temperature oxidation, then DON could be measured directly and DIN by difference (DIN = TDN – DON). This work leads to an important hypothesis for this study: Removing DIN from water samples prior to analysis decreases the MDL for DON in drinking and reclaimed wastewaters.

DON Occurrence

Analysis of data reported by the United States Geologic Survey (USGS) and the USGS National Water Quality Assessment (NAWQA) program for approximately 23,000 surface water and shallow or deep groundwater samples, was used to estimate DON occurrence in USA surface waters (Westerhoff et al. 2002). The USGS employed TKN measurements and ammonia measurements, from which DON was calculated by difference:

$$\text{DON} = \text{TKN} - \text{Ammonia} \tag{1.3}$$

Approximately 30% of the surface water sites consisted of measurements that were reported to be less than the TKN detection limit of 0.2 mgN/L, and hence could not be statistically considered. Seventy-six percent (76%) of the shallow groundwater beneath specific land uses and 81% of the deep groundwater aquifer samples also had TKN values less than 0.2 mgN/L, suggesting that groundwaters typically contain low DON concentrations (Nolan and Stoner 2000). Bar and whisker diagrams for DON occurrence in surface waters with detectable TKN values are presented in Figure 1.1. The median DON concentration was 0.37 mgN/L, but would be lower if the 30% of samples with TKN less than 0.2 mgN/L were considered. The DOC/DON ratios for the same data are also plotted in Figure 1.1, and indicate a median ratio value of 15 mgC/mgN. A low DOC/DON ratio is probably indicative of more autochthonous DOM sources, while a DOM with higher DOC/DON ratios represents allochthonous DOM sources (Aiken and Cotsaris 1995). This data provides an excellent baseline for DON occurrence in potential drinking water supplies, and analysis of watershed hydrology could provide further insight. Furthermore, DOC/DON ratios may serve as potential indicators of DOM origin.

Organic nitrogen and DBP Formation

DON can lead to the formation of nitrogenous DBPs of health and regulatory concern (e.g., nitrosamines, haloacetonitriles (HAN), nitromethanes, cyanogen halides) (Klotz and Pyrch 1999; Krasner et al. 2001; Richardson et al. 1999). Using comparative genotoxicity and cytotoxicity assays, it was concluded that nitromethanes, especial di- and tri- brominated nitromethanes, have high potential human toxicity; brominated acetic acid was also highly genotoxic (Plewa et al. 2000; Plewa et al. 2002). While the drinking water industry has recognized the significance of DOC and bromide as critical precursors for DBP formation, additional recognition, quantification and understanding of the role of DON as critical precursor material is necessary.

PROJECT OBJECTIVES

The goal of this first AwwaRF project on dissolved organic nitrogen (DON) is to document DON occurrence and provide information on its chemical characteristics and reactivity towards metal hydroxides and oxidants/disinfectants. Relatively minor attention has focused on DON despite observations that both low and high molecular weight molecules containing organic nitrogen (e.g., simple amino acids, algal-derived humic substances) have been implicated as disinfection by-product (DBP) precursors (trihalomethanes (THMs), haloacetic acids (HAAs), haloacetonitriles (HANs), nitromethanes, nitrosamines), and membrane foulant materials. Furthermore, while higher organic nitrogen content is associated with increasingly polar natural organic matter (NOM) fractions, there are no publications of DON removal during coagulation. The proposed research will use both well-developed DON analytical methods from the fields of ecology and marine biology and innovative DON quantification methods tailored for the drinking and reclaimed wastewater industry to meet the following specific objectives:

- Develop an accurate DON quantification method suitable for drinking waters
- Develop a comprehensive assessment of DON occurrence in raw/finished drinking waters and reclaimed wastewaters
- Evaluate a suite of analytical tools for characterizing DON structure
- Quantify DON removal by sorption processes (coagulation, activated carbon)
- Investigate role of DON on DBP formation (HAN, NDMA, THM) during disinfection
- Validate methods for organic chloramine measurement and assess organic chloramine formation during disinfection (free chlorine and monochloramine)
- Assess future organic-nitrogen research needs.

Median
The data value located halfway between the smallest and largest values.

Upper Quartile (UQ)
The data value located halfway between the Median and the highest data value.

Lower Quartile (LQ)
The data value located halfway between the Median and the lowest data value.

Interquartile Distance (IQD)
The distance between the Upper and Lower Quartiles (UQ - LQ).

Outliers
Points whose value is either:
greater than UQ + 1.5 * IQD
or
less than LQ - 1.5 * IQD

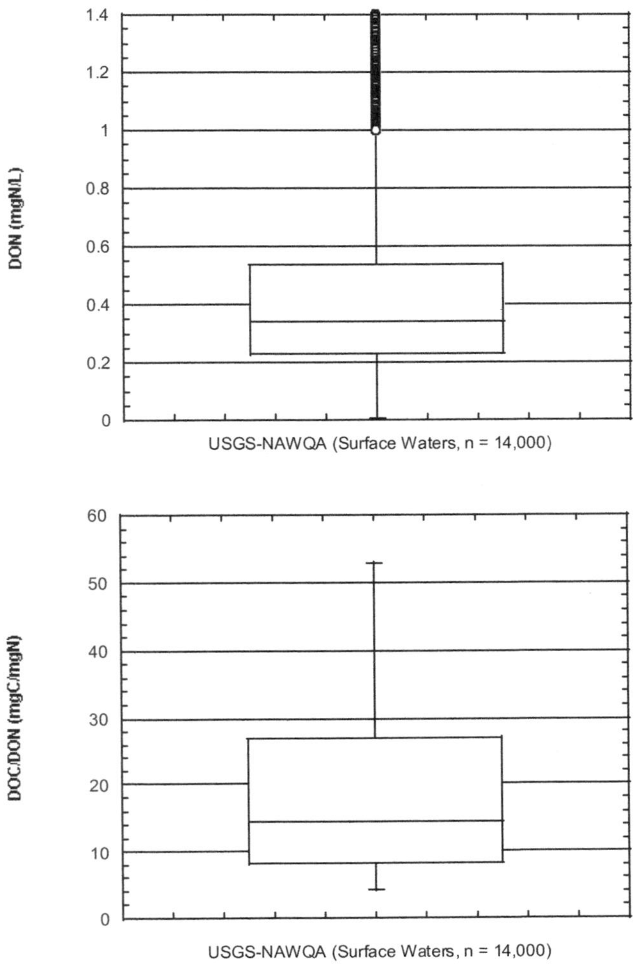

Source: Based on data from (Westerhoff and Mash 2002).

Figure 1.1 Box plots for DON occurrence (upper) and DOC/DON ratios (lower).

CHAPTER 2 MATERIALS AND METHODS

This chapter describes sampling locations and methods, experimental methods for laboratory tests, and analytical methods. Chapter 3 details the development, verification, and application of a new dialysis-based DIN pretreatment method aimed at improving the accuracy of DON analysis.

SAMPLING LOCATIONS & METHODS

Utility Survey Sampling

Twenty eight water treatment plants (WTPs) in 13 states (Arizona, California, Illinois, Indiana, Kentucky, Massachusetts, Michigan, Missouri, Nevada, Pennsylvania, Texas, Virginia, and Wisconsin) participated in the two sampling campaigns. Figure 2.1 illustrates the geographical distribution of these WTPs. Table 2.1 summarizes individual utility names, locations and reference numbers. Raw water from these facilities include rivers (21 WTPs), reservoirs (5 WTPs), and lakes (2 WTPs).

The first sampling campaign was conducted during late spring and early summer (May or June) of 2003, and the second campaign during fall or winter (October, November, or December) of 2003. Two utility specific sampling campaigns were also conducted to study shorter-term temporal variability in DON at the following locations: 1) El Capitan Reservoir, Calif. once per week for five weeks (June through July 2003), and 2) Muncie, Ind. once per month from February through July 2004. Sampling kits placed in small coolers were mailed to each facility. The kits included sampling instructions, pre-cleaned 1 L amber glass bottles with Teflon coated caps, a disposable syringe, and a small vial of sodium thiosulfite (50 mM Na_2SO_3). Utility staff were instructed to collect duplicate grab samples of raw and finished water. Exactly 1 mL of sodium thiosulfite was added to all finished water samples to quench the disinfectant residual. Collected samples were bubble-wrapped, placed in the small coolers, ice-pack added, and shipped overnight to Arizona State University. Upon receipt samples were filtered and refrigerated (4 ^{o}C) and analyzed within seven days.

Additional Reservoir Sampling

Central Arizona

Surface water samples were collected from two canals (Central Arizona Project, CAP, and Arizona canals) and three reservoirs (Lake Pleasant, Bartlett, and Saguaro Lakes) supplying drinking water to Phoenix's Metropolitan Area. Lake Pleasant, Bartlett, and Saguaro Lakes are reservoirs of the CAP, Verde River and Salt River systems, respectively (Figure 2.2); sampling sites are identified by the R# designation. Details about this watershed are available elsewhere (Nguyen et al. 2002). Reservoir samples were collected from the 10 m and ~ 30 m depths using a Kemmerer sampling device, from locations near the outlets of each reservoir. Raw and treated water samples were collected from three water treatment plants (WTPs) from the City of Phoenix (Deer Valley (R16, R17), Squaw Peak (R14, R15) and Val Vista (R18, R19) WTPs). Depending on the season, these WTPs use a mixture of surface water from three sources: CAP, Verde River and Salt River systems. Water is conveyed from the upstream sources to the WTPs via open-channel, concrete-lined canals; groundwater pumping and irrigation returns add water into the canal along its length. Samples were collected in pre-cleaned amber glass bottles and stored in a cooler at 4 °C during their transport to the laboratory.

Lake Mead, Nevada

All sampling and analysis on Lake Mead, on the Colorado River, was conducted by Southern Nevada Water Authority (Figure 2.3). Las Vegas Wash conveys largely discharged effluent and stormwater runoff into Las Vegas Bay (LVB). Southern Nevada Water Authority has a raw water intake structure east of Saddle Island. Electronic data from multiple water depths at five sampling locations was provided to our research team. Total nitrogen, ammonia, and nitrate data was provided. DON was calculated by subtracting ammonia and nitrate concentrations (as μgN/L) from total nitrogen.

South Platte River, Colorado

The University of Colorado (Boulder) collected samples from the South Platte River system (Figure 2.4). The South Platte River along the Front Range in Colorado receives effluent discharges from 25 wastewater treatment plants, making up 95% of the total daily wastewater effluent discharged into the entire South Platte River Basin (Dennehy et al. 1998). These discharges can comprise a substantial portion of the streamflow. The largest single wastewater effluent discharged into the South Platte River comes from the Metro Wastewater Reclamation District Treatment Plant (Metro WWTP). This discharge makes up 69% of the streamflow annually. During extreme scenarios, it makes up 100% of the streamflow in the river. Extreme scenarios describe days of no river flow upstream of the Metro WWTP. At a minimum, the South Platte is wastewater-impacted and, at a maximum, wastewater-dominated. A municipality using the South Platte River as a drinking water source is indirectly and unintentionally reusing wastewater effluent.

The sampling points were chosen so that water downstream of wastewater discharges could be compared with water upstream. The two principal wastewater discharges along the study segment were the Littleton/Englewood Wastewater Treatment Plant (Littleton/Englewood WWTP) and the Metro WWTP. Other plants discharge along the study segment, but they were not considered because their effluent flows are insignificant compared to the effluent flows from the Littleton/Englewood WWTP and the Metro WWTP. As it is, the Littleton/Englewood WWTP releases a much smaller average discharge, 37 cubic feet per second (cfs), compared to the Metro WWTPs average of 230 cfs. One sample from the Littleton/Englewood WWTPs effluent and two different samples from the Metro WWTPs effluent were collected. In addition, five South Platte River samples were collected upstream and downstream of these two discharges, totaling eight samples. Their descriptions and specific sampling locations are explained below.

Two sets of samples were collected at their respective locations for this investigation. The first set of samples was collected on April 25, 2003. The second set was collected on May 14, 2003. On both days, all eight samples were collected within a four-hour period. All samples were collected in one-gallon plastic jugs. The Chat R and SP at 285 samples were collected by wading near the water's edge. The samples L/Eeff, SP at Burl, M-2eff, M-N/Deff, SP at SC, and SP at Thor were collected with a bucket suspended by a rope or rod. All samples were transported by car to the University of Colorado at Boulder, where they were 0.45-μm filtered into amber vials and stored at 4 °C in the dark.

BENCH-SCALE TESTING

Coagulation Jar Tests

Five raw waters from participating utilities (Harwoods Mill reservoir, Virginia; Huron River, Michigan) and reservoirs in Arizona (Lake Pleasant, Bartlett Lake, Saguaro Lake) were collected in 20 L Nalgene high-density polyethylene containers and used in coagulation or activated carbon adsorption experiments. Coagulation experiments used a jar test apparatus (Stirrer Model 7790-400, Phipps and Bird, Richmond, VA). Each sample (1.5 L) was rapidly mixed at 100 revolutions per minute (rpm) for 2 min, followed by a slow-mixing at 30 rpm for 20 min, and then settled (0 rpm) for 60 min. Aluminum sulfate ($Al_2O_{12}S_3 \cdot 18H_2O$) and a cationic polymer (CLARIFLOC C-308P (polyDADMAC), POLYDYNE Inc., GA) were added as a coagulant (0 to 15 mg/mgDOC) and a coagulant aid (0, 1, 2 and 3 mg/mgDOC), respectively, during the rapid mixing. The organic nitrogen content of the polymer was 7-8% (measured). At the end of each experiment, samples were filtered (Whatman GF/F).

Activated Carbon Adsorption Experiments

For activated carbon adsorption experiments the same water sources as for the above coagulation experiments were used. However, the water was filtered (GF/F) prior to use. PAC adsorption experiments were conducted using local surface water from Central Arizona Project (CAP) canal, which was spiked with a hydrophobic NOM isolate from CAP water, and a tertiary treated denitrified wastewater effluent. The effluent was collected from the Northeast Mesa Wastewater Treatment Plant (WWTP) (Mesa, AZ), which is an activated sludge process (SRT = 10 day) that incorporates nitrogen removal through anoxic (HRT = 1.5 h) and aeration zones (HRT = 3 h). Samples were immediately filtered using ashed (500 ^{o}C) glass fiber filters (Whatman GF/F, nominal pore size = 0.7 μm).

Batch PAC sorption tests were conducted using various types of PACs (Table 2.2): AC800 (AC800, Acticarb, Dunnellon, FL, USA), Hydrodarco-B #1033 (HDB-1033), Hydrodarco-B #1091 (HDB-1091), PAC 20B (Norit Americas Inc., Atlanta, GA, USA), Picazine (Pica, Pica, France), and WPM (PAC form of F400, Calgon Carbon Corp., Pittsburgh, PA, USA). Applied PAC doses for screening experiment were 5 and 25 mg/L. Batch reactors were 40 mL amber glass vials and were shaken on a shaker for 20 h. After PAC adsorption experiment, samples were filtered using ashed GF/F prior to analysis.

Organic Chloramine Formation Experiments

Organic chloramine kinetics and yields were assessed through free chlorine or preformed monochloramine addition to solutions prepared using NOM isolates or dialyzed raw water samples. For solutions containing NOM isolates, most experiments were conducted with 2 mM of a phosphate buffer at pH 7.5 and 20 ^{o}C. Samples from the utility survey were also chlorinated. However, in order to remove ammonia and other salts before chlorine, the raw water samples were filtered (GF/F) and then subjected to dialysis. After dialysis all samples were adjusted to pH 7.5 and 2 mM of phosphate buffer added. In all cases the initial solutions contained ammonia concentrations below detection limits (< 5 μgN/L).

Hypochlorite solutions (Fisher Scientific) were diluted and used to prepare free chlorine and preformed monochloramine stock solutions. Monochloramine stock solutions were prepared at a concentration of ~ 250 $mgCl_2/L$ by mixing hypochlorite and ammonia chloride solutions (pre-adjusted to pH 8.4) at a ratio of 0.8 mol/mol. The monochloramine stock solution was refrigerated for at least 3 hours before use, and was prepared daily.

Experiments were conducted either in 40 mL amber glass bottles with gas-tight septa caps or 1 L completely stirred repipette amber glass containers. Samples were collected for total chlorine, free chlorine, and monochloramine measurement. Monochloramine was only measured in initial experiments and those containing preformed monochloramine. The organic chloramine concentration ($mgCl_2/L$) was calculated by difference:

$$[\text{Organic Chloramine}] = [\text{Total Chlorine}] - [\text{Free Chlorine}] - [\text{Monochloramine}] \quad (2.1)$$

Disinfection By-Product Formation Experiments

Free chlorine or preformed monochloramine was added to solutions prepared using NOM isolates to study the formation of DBPs. Samples containing 3 mg/L of DOC were prepared at pH 7 (5 to 10 mM phosphate buffer) which resulted in solutions with a range of DON concentrations (50 to 500 μgN/L) since the nitrogen content varied in the NOM isolates (Table 2.3). See above section for preparation of free chlorine and preformed monochloramine solutions. All experiments were conducted in 1 L amber glass bottles with gas-tight Teflon caps, and incubated in the darkness at 20 °C. Free chlorine DBP formation experiments were conducted following the approach of (Reckhow et al. 1990), using a dose of 5 $mgCl_2$ per mg DOC with a reaction time of 7 days. Monochloramine DBP formation experiments were conducted following the approach of (Mitch and Sedlak 2004), using a dose of 45 $mgCl_2$ per mg DOC with a reaction time of 10 days. Oxidant residuals were quenched at the end of the reaction time, and DBPs analyzed by the City of Phoenix Lab Services using USEPA Method 551.1 for THMs and HAN or Montgomery Laboratories using USEPA Method 525s for NDMA.

MOLECULAR WEIGHT FRACTIONATION

Two complimentary approaches were used to evaluate molecular weight distributions of DOC or DON. First, size exclusion chromatography with in-line UV, fluorescence, and DOC detection was used for samples collected from the South Platte River system. Details of instrument setup, calibration, and data interpretation are provided elsewhere (Her et al. 2002a; Her et al. 2002b; Her et al. 2003). Second, molecular weight distributions of DOC, UVA_{254} and DON were determined using 400-ml ultrafiltration cells (Model 8400, Amicon, Beverly, MA) pressurized to 380 KPa with nitrogen gas. Ultrafiltration membranes with nominal molecular weight cutoffs of 1,000 Daltons (Amicon YM1) or 10,000 Daltons (Amicon PM10) (Esparza-Soto and Westerhoff 2001).

GLASSWARE, FILTERS, AND REAGENTS

All glassware was washed with Alconox soap, rinsed with tap and distilled water, acid washed (10% HCl) and ashed at 550 °C (overnight) prior to use. All filtration was conducted using Whatman (Millipore Corp., MA) glass fiber (GF/F) filters which have a nominal pore size of 0.7 μm. All filters were ashed at 550 °C prior to use. Approximately 50 mL of Nanopure water (NANOpure Infinity, Barnstead, IA) and then 50 mL of sample were used to rinse the filters prior to processing the sample. All chemicals were reagent grade or better. Standards were prepared in Nanopure water.

NATURAL ORGANIC MATTER ISOLATES

From previous AwwaRF projects and related studies J.P. Croue from the University of Poitiers isolated and characterized a wide range of NOM isolates. The University of Poitiers isolates and others obtained from the International Humic Substances Society (St. Paul, MN) were used in different parts of this current study. Solutions containing the NOM isolates were prepared by dissolving the isolates in Nanopure water, sonicating the sample, and filtering the sample (GF/F) prior to use. Methods detailing the NOM isolation procedures and characterization of the NOM isolates are provided elsewhere (Croue et al. 1999). The source, type of NOM fraction, DOC/DON ratio, and SUVA values for the NOM isolates are summarized in Table 2.3. The DOC/DON ratios of the NOM isolates ranged from 5 to 60 mgC/mgN, and SUVA values from 1 to 5 L/mg-m.

ANALYTICAL METHODS

Table 2.4 summarizes most of the methods and instrumentation used for this study. Other methods are described below. Of particular interest in this study was nitrogen-based measurement techniques, which are reviewed elsewhere (Westerhoff and Mash 2002). Of particular interest was determination of minimum detection limits (MDLs) and method comparisons for different sources of nitrogen.

MDLs were determined by multiplying the standard deviation between the replicates for each species by the t-value at 99% confidence and n-1 degrees of freedom (USEPA 1999). MDL is defined as the minimum concentration of a substance that can be measured and reported with 99% confidence, and was determined based upon the USEPA protocol (USEPA 1999). Briefly, the MDL was determined by multiplying the standard deviation between the replicates for each species by the t-value at 99% confidence and n-1 degrees of freedom. Solutions were prepared containing 0.01, 0.02, and 0.05 mg/L of each model compounds in deionized water, and seven replicates were analyzed using the methodology. Independent or paired t-test was used based on normality to compare two groups of data. In case the data were not normally distributed and did not have equal variances, the Wilcoxon signed-rank test was used. Statistical analysis of experimental data was performed using SPSS V. 11.0 (SPSS Inc., IL, U.S.A.). MDLs were calculated for several model compounds (Table 2.5). High temperature oxidation (HTO) had equal or lower MDLs than wet chemical oxidation (WCO) methods for 14 model compounds raining from 0.09 to 0.019 mgN/L, and were comparable to other analytical methods for inorganic nitrogen determination (Ammann et al. 2000; Sharp et al. 2002).

Several types of TDN standards were compared: 100% nitrate, 50% nitrate/50% ammonia, 100% glycine, and mixtures of 17 amino acids. Calibration curves had nearly identical slopes and linear regression coefficients. Reproducibility tests were conducted using solutions of glycine and a solution of NOM isolate (Ribou Reservoir HPO fraction). The relative standard deviations calculated on 5 and 20 replicates for glycine and NOM solutions respectively are reported in Table 2.6. One vial was used per replicate since previous analyses have shown high RSD values when injecting several times the same solution (several injections from the same vial). The results show that the TN analyzer is reproducible with RSD below 6% for glycine concentrations ranging from 20 to 1000 μg N/L. However, the mean concentrations calculated based on the calibration curve established from the mean area values are relatively similar for the two lowest concentrations, 23 and 29 for 20 and 50 μg/L, respectively. This observation seems to indicate that the quantification limit for this TDN analyzer is above 50 μg/L. Results obtained for NOM solutions prepared with the HPO fraction isolated from the Ribou Reservoir were found to be in accordance with the concentrations calculated based on its elemental analysis (N content was obtained from N_2 analysis using thermal conductivity detection after NOM combustion at 1050 °C). The relative standard deviation was found to be 2.8 % for an organic nitrogen content approaching 300 μg DON/L.

To compare the nitrogen analyzers, 8 solutions of NOM isolates (TPI and HPO fractions) were analyzed with both ThermoEuroglass 3000 and Shimadzu TNM-1 analyzer. Glycin was used as a standard for both analyzers. Figure 2.5 shows the analyzed nitrogen concentrations vs. the nitrogen concentration calculated from elemental analysis. Results show that for 5 NOM fractions Shimadzu analyzer gave TN values higher than values obtained with TN 3000 but also higher than TN concentrations calculated from elemental analysis. Figure 2.5 shows also that standard deviations are much lower with Shimadzu analyzer.

High temperature oxidation (HTO) and wet chemical oxidation (WCO) TDN methods were directly compared using 110 surface water samples. The results indicated no difference between the two methods (significance level, $\alpha = 0.05$) (Figure 2.6). The HTO method was also checked for nitrogen recovery with various nitrogen-containing compounds (1 mgN/L), and the recoveries were 97–103% depending on the compounds. The HTO method was simpler (autosampler rather than manual digestion) and yielded both TDN and DOC concentrations with the Shimadzu TOC-V instrument, therefore it was used for most analysis in this report, unless specifically stated that WTO methods were employed. Details of DON measurement are presented in Chapter 3, but DON concentration was calculated by the difference of TDN and DIN:

$$DON = TDN - NO_3^- - NO_2^- - NH_3/NH_4^+ \qquad (2.2)$$

Amino Acids and Sugar Analysis

Total amino acids and total aminosugars were analyzed by HPLC after acid hydrolysis (6 N HCl, 6 hours) and precolumn derivatization using the AQC derivatization agent (6-aminoquinolyl-N-hydroxysuccinimidyl carbamate):

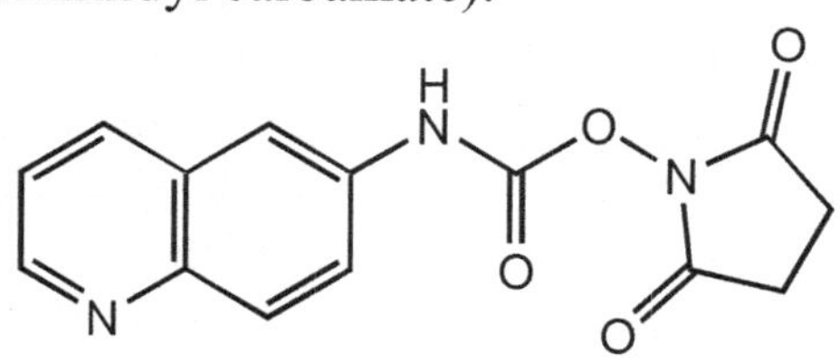

6-aminoquinolyl-N-hydroxysuccinimidyl carbamate

The reaction between AQC and amino acids (primary and secondary amino acids) and aminosugars leads to the formation of highly fluorescent complexes that can be analyzed using fluorescence emission detection:

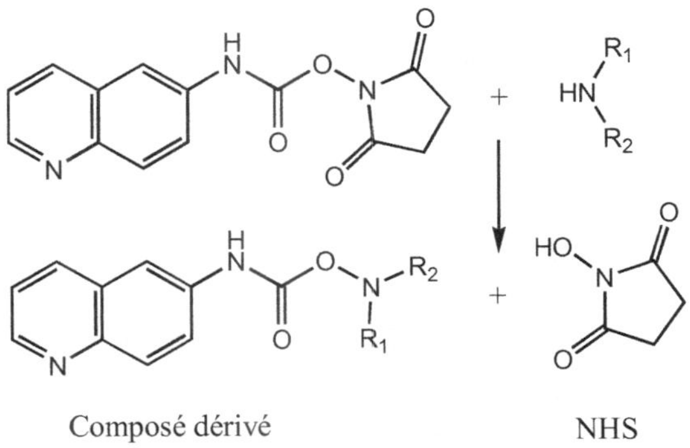

Derivatization reaction

During the derivatization reaction, three other byproducts are also produced 6-aminoquinoline (AMQ), N-hydroxysuccinimide (NHS) and carbon dioxide (below). These three by-products do not interfere with the rest of the chromatogram.

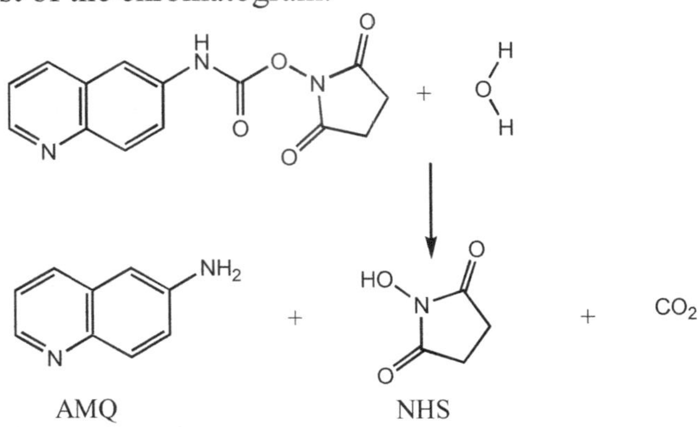

By-products formation during AQC derivatization

Instrumentation comprised a Waters 600 E solvent pump, a Merck AS 4000 autosampler and a Merck F-1050 fluorescence detector. Data acquisition was conducted using the Waters Millennium software. The acid hydrolysis is operated using the WATERS Pico.Tag unit. 100 µL of sample are introduced in a tube (6 x 50 mm). A maximum of 12 tubes are placed in a reaction vial. After dryness operated under vacuum, 200 µL of 6N HCL are introduced into the reaction vial with the addition of cristal of phenol. The acid hydrolysis is conducted at 116 °C under nitrogen atmosphere during 6 hours. After hydrolysis the residual of HCL is evaporated under vacuum followed by the addition in each tube of 20 µL of the internal standard solution, 60 µl of a borate buffer and 20 µL of the AQC solution. The tube is then introduced in an autosampler vial. 10 µL of the final solution are injected for analysis.

After derivatization, aminoacids and aminosugars are separated on a AccQ.Tag (3,9 mm x 150 mm, C_{18}) HPLC column in the presence of α-aminobutyric acid as internal standard using an solvent elution program that use three different eluents (Eluent A that is a mixture of sodium acetate, triethyl amine, EDTA phosphoric acid and sodium azide ; acetonitrile and MilliQ water). The elution program proposed by Waters is detailed in Table 2.7.

After separation, amino acids and aminosugars are detected using a fluorescence detector with 240 nm and 395 nm as excitation and emission wavelengths, respectively. Figure 2.7 shows a typical chromatogram of a standard solution made of 17 amino acids and 2 aminosugars (amino acid and aminosugar concentrations of 20 µM except for cystine 10 µM). Table 2.8

gives the relative standard deviation of the analysis of 4 replicates of a mixed standard solution of amino acids and aminosugars after AQC derivatization. Table 2.9 gives the slope and the linear regression coefficient R^2 of the calibration curves established for the 17 amino acids and the 2 aminosugars studied for concentrations ranging from 0 to 10 μg/L.

Nitrogen Nuclear Magnetic Resonance

Spectra were acquired on a Varian CMX Infinity 300 MHz solid state NMR spectrometer, equipped with a wide-bore magnet and a probe for double and triple resonance studies. The NMR frequency for ^{15}N is in the vicinity of 30.4 MHz at this magnetic field (~7 Tesla). A careful check with ^{13}C CP/MAS was performed, to assure that the system's performance was within manufacturer's specifications. Standard samples with enrichment to 98% in ^{15}N were used to optimize operating conditions. Samples were spun at 3.3 kHz, and a standard single-decoupling H pulse program was used, with contact time 5 ms and H r.f. amplitude approximately 1.5 mT. The working chemical shift reference compound employed for solid state spectra was initially $Ca(^{15}NO_3)$, but we have recently begun to use $^{15}NH_4{}^{15}NO_3$, as recommended by some researchers (MacKenzie and Smith 2002). Values for chemical shift are given herein, however, in terms of the primary reference for ^{15}N spectra, namely nitromethane (CH_3NO_2). The conversion factor between nitromethane and the NH_4^+ reference line of $^{15}NH_4{}^{15}NO_3$ was obtained and is -358.0 ppm (Turner et al. 1987). That is, to convert chemical shift values referenced to $^{15}NH_4^+$ in solid $^{15}NH_4{}^{15}NO_3$ to nitromethane-referenced values, one simply subtracts 358.0. This results in most shifts being negative, because the nitrogen in nitromethane is rather strongly deshielded, but the reproducibility of results obtained using this compound outweigh the inconvenient feature of sign, and have resulted in its adoption as the primary standard for ^{15}N NMR spectra in both liquids and solids.

CP/MAS Spectra of ^{15}N are considerably more difficult to acquire than those of ^{13}C, owing in part to ^{15}N's lower natural abundance but also frequently to special features of ^{15}N's nuclear resonance properties, for example its very long spin relaxation times, narrow lines and negative gyromagnetic ratio (this last leading to negative values of the Nuclear Overhauser Effect, in solids as well as liquids) (Levy and Lichter 1979). The resulting variations in signal-to-noise ratios can be extreme, even between chemical species with similar numbers of nitrogens per unit volume. An example of such variation is seen in our spectra for glycine vs. those of uracil. The signal-to-noise ratio for glycine, with only one N per molecule, is approximately four times better after 100 data acquisitions and with 100 Hz line broadening, than for uracil with two Ns per molecule after 30,000 acquisitions and 200 Hz line broadening. Possibly the parameters for uracil acquisition could be improved via minor adjustments or changes in pulse scheme, since the sensitivity of CP/MAS spectra to spectrometer settings is notably large, but this ~ 60-fold difference in signal amplitudes points to large differences in ^{15}N line widths between the two compounds. In general, weak signals and long acquisition times are normal in working with this nucleus. In the soil chemistry literature accounts of acquisitions of up to 1 million sweeps may be found, especially for complex NMR experiments. Soil samples are for this reason frequently treated with HF, in order to concentrate organic matter (Kogel-Knabner 1997).

Table 2.1 Reference numbers, name, and location of facilities participating in utility sampling campaigns

Reference Number for Utility	Facility Name	Facility Location
1	Gilbert WTP	Gilbert, AZ
2	Greenway WTP	Peoria, AZ
3	24th Street WTP	Phoenix, AZ
4	Deer Valley WTP	Phoenix, AZ
5	Val Vista WTP	Phoenix, AZ
6	Union Hills WTP	Phoenix, AZ
7	Bollman Treatment Plant	Concord CA
8	Central Lake County Joint Action Water Agency	Lake Bluff, IL
9	Alton WTP	Alton, IL
10	East St. Louis WTP	E. St. Louis, IL
11	Granite City WTP	Granite, IL
12	Peoria District WTP	Peoria, IL
13	Streator WTP	Streator, IL
14	Pontiac WTP	Pontiac, IL
15	Kokomo Operations	Kokomo, IN
16	Richmond Middle Fork WTP	Richmond, IN
17	Muncie Operations	Muncie, IN
18	Louisville Water Company	Louisville KY
19	MWRA	Southborough, MA
20	Ann Arbor Water Plant	Ann Arbor, MI
21	Joplin WTP (MAWC)	Joplin, MO
22	Jefferson City WTP	Jefferson City, MO
23	St. Louis County WTP	Chesterfield, MO
24	Southern Nevada Water Authority	Boulder City, NV
25	Philadelphia WTP	Philadelphia PA
26	Eagle Mt. WTP	Fort Worth, TX
27	Newport News Waterworks	Yorktown, VA
28	Milwaukee Water Works	Milwaukee, WI

Table 2.2 Characteristics of PAC brands

Properties	AC800	HDB-1091	HDB-1033	Norit 20B
Iodine no. (mg/g)	800	540	540	800
Molasses no.	N/A	500	500	190
Material type	Coal	Lignite coal	Lignite coal	Bituminous coal
Tannin value (mg/L)	N/A	220	220	900

Table 2.3 DOC/DON ratios and SUVA values of NOM isolate solutions

Source	Type of Fraction	DOC/DON (mgC/mgN)	SUVA (L/mg-m)
Seine River Raw	HPO	15	3.0
Ribou Reservoir Raw	Colloids	10	3.7
Ribou Reservoir (Clarified + PAC)	HPO	21	2.3
Ribou Reservoir (Clarified + PAC)	TPI	9	1.3
South Platte River Raw	HPO-A	26	3.0
South Platte River Raw	TPI-N	6	0.9
Suwannee River	TPI-N	28	2.8
Naintre Effluent Organic Matter	Colloids	9	4.6
Naintre EfOM	HPO	14	2.4
Colorado River, Biofilter Effluent	HPO	36	2.2
Colorado River, Biofilter Effluent	TPI	15	1.0
Suwannee River	TPI-A	40	3.5
Suwannee River	HPIA+N	28	2.7
Nordic Reference	FA	58	5.0
Nordic Reference	TPI-A	45	4.0
Britanny River	TPI	11	1.5
Britanny River	HPO	19	3.4
Britanny River	Colloids	11	2.9

Note: HPO = hydrophobic; HPO-A = hydrophobic acid ; TPI = Transphilic : TPI-N = Transphilic neutral ; TPI-A = transphilic acid ; HPIA+N = Hydrophilic acid plus neutral.

Table 2.4 Summary of analytical methods and instruments

Parameter	Method Description	Instrument Description	Method or Reference
Dissolved organic carbon	High temperature combustion	TOC-V_{CSH} with TNM-1 TN unit (Shimadzu, Japan)	5310B
Total dissolved nitrogen	High temperature combustion (720 °C) with chemiluminescence detection		(Sharp et al. 2002)
Total dissolved nitrogen	High temperature combustion (900 °C)	Euroglass TN3000	
Total dissolved nitrogen	Alkaline persulfate oxidation and automated cadmium reduction using a continuous flow wet chemistry analyzer	TrAAcs 800 AutoAnalyzer (Bran-Luebbe, Germany)	4500-N C 4500-NO_3^- F
Ammonium	Automated phenate method		4500-NH_3 G
Nitrate	Automated cadmium reduction method		4500-NO_3^- F
Nitrite			4500-NO_2^- B
Ammonium*	Ion Chromatography with	Alltech SHMIXR 100x4.6mm, 70% 4 mM HNO_3/30% methanol.	
Nitrate & Nitrite	Ion Chromatography with conductivity detection	DX-120 (Dionex Corp., CA)	4500-NO_3^- C
Nitrate & Nitrite*	Ion Chromatography with UV (215 nm) detection	C18 column (5 µm: 250x4.65 mm), 0.01 M KH_2PO_4 buffer (pH 6), 2.5 mM tetrabutyl ammonium sulfate.	
Free Chlorine	Free DPD method	DR2000 (Hach, Loveland, CO)	4500-Cl G
Total Chlorine	Total DPD method		4500-Cl G
Monochloramine	Indophenol method	MonoChlorF (Hach, Loveland, CO) UV 1601 diode-array spectrophotometer (Shimadzu, Japan)	Hach
Proteins	Lowry Method standardized with bovine serum albumin (BSA) standard	Bio-Rad Protein Assay Kit (Bio-Rad Laboratories, CA)	(Lowry et al. 1951)
Proteins	Fluorescence binding reagent standardized with bovine serum albumin (BSA) standard	NanoOrange Protein Quantitation Kit (Molecular Probes Inc., OR) LS50B Fluorimeter (Shimadzu, Japan)	(Jones et al. 2003)
pH	Gel electrode	Model 511201 (Beckman pH meter)	4500-H^+ B
Conductance	Conductivity meter	Model 2052 (VWR conductivity meter)	2510 B

*Used by University of Poitiers. MDL for nitrate, nitrite, and ammonium were 0.002, 0.002, and 0.020 mgN/L.

Table 2.5 MDLs for nitrogen containing compounds by different analytical methods

Compound	Molecular formula	MDL (mgN/L)			
		HTO	WCO	IC	TrAAcs
Inorganic nitrogen					
Potassium nitrate	KNO_3	0.009	-	0.005	0.013
Sodium nitrite	$NaNO_2$	0.009	-	0.005	0.014
Ammonium chloride	NH_4Cl	0.010	-	0.005	0.005
Amino acids					
Arginine	$C_6H_{14}N_4O_2$	0.007	0.033	-	-
Glutamic acid	$C_5H_9NO_4$	0.013	0.019	-	-
Glycine	$C_2H_5NO_2$	0.016	0.010	-	-
Proline	$C_5H_9NO_2$	0.010	0.022	-	-
Tryptophan	$C_{11}H_{12}N_2O_2$	0.009	0.032	-	-
Nucleic acids and proteins					
RNA		0.018	0.017	-	-
Bovine serum albumin		0.019	0.018	-	-
Tri-peptide (Glu-Cys-Gly)	$C_{10}H_{17}N_3O_6S$	0.014	0.020	-	-
Other org-N compounds					
Aniline	C_6H_7N	0.014	0.013	-	-
Imidazole	$C_3H_4N_2$	0.011	0.011	-	-
Urea	CH_4N_2O	0.013	0.019	-	-

Table 2.6 TDN analysis: reproducibility tests for glycine and NOM solutions

Standard	Concentration	TDN signal			TDN (μgN/L)†	
	μg N/L	Mean Area	Standard deviation	RSD %	Mean value	Standard deviation
Glycine	20	356	± 11.99	3.37	22.78	± 4.26
Glycine	50	372	± 20.05	5.7	29.2	± 7.13
Glycine	100	661	± 31.11	4.71	134	± 7.75
Glycine	500	1651	± 24.53	1.48	484	± 8.63
Glycine	1000	3118	± 21.78	0.70	1006	± 5.98
HPO NOM	1240 *	3829	± 62.76	1.64	1288	± 22.4
Ribou Reservoir	248 *	946	± 20.87	2.20	287	± 7.94

* Calculated based on the elemental analysis : 50 mg NOM /L with 52.5 % C and 2.48 % N)
† Recalculated based on the calibration curve obtained from the mean area value N (μg/L) = 0.3555 (Area) – 103.71 ; $R^2 = 0.9975$)

Table 2.7 HPLC elution program for the amino acid and aminosugars analysis

Time (min)	Flow (mL/min)	Eluent A	Acetonitrile	MilliQ	Profile (Waters pump)
Initial	1.0	100	0	0	*
0.5	1.0	99	1	0	11
18.0	1.0	95	5	0	6
19.0	1.0	91	9	0	6
29.5	1.0	83	17	0	6
33.0	1.0	0	60	40	11
36.0	1.0	100	0	0	11
65.0	1.0	0	60	40	11
100.0	0.0	0	60	40	6

Table 2.8 Relative standard deviation of the analysis of 4 replicates of a standard amino acid and aminosugar solution

Standard	Conc μM	RSD %	Compound	Conc μM	RSD %
Aspartic acid (Asp)	20	4.5	Cystine (Cys)	10	4.1
Serine (Ser)	20	4.2	Tyrosine (Tyr)	20	4.3
Glutamic acid (Glu)	20	4.1	Valine (Val)	20	3.9
Glycine (Gly)	20	4.5	Methionine (Met)	20	4.5
Histidine (His)	20	4.5	Lysine (Lys)	20	4.9
Ammonia (NH_3)	20	10.9	Isoleucine (Ile)	20	4.3
Arginine (Arg)	20	4.8	Leucine (Leu)	20	5.0
Threonine (Thr)	20	4.2	Phenylalanine (Phe)	20	1.5
Alanine (Ala)	20	5.2	Galactosamine	20	4.3
Proline (Pro)	20	4.1	Glucosamine	20	4.5

Table 2.9 Calibration curve parameters (0 to 10 μg/L range) for specific amino acids and aminosugars

Compound	Slope (x 10^{-5})	Linear regression coef R^2
Asp	2.81	0.9997
Ser	5.10	0.9996
Glu	3.17	0.9997
Gly	3.44	0.9996
His	5.38	0.9998
Arg	5.68	1.0000
Thr	5.53	0.9998
Ala	6.43	0.9992
Pro	3.25	0.9997
Cys	69.4	0.9997
Tyr	5.68	0.9998
Val	9.90	0.9998
Met	7.85	0.9998
Lys	3.90	1.0000
Ile	11.35	0.9998
Leu	10.20	0.9998
Phe	4.125	0.9802
Galactosamine	1.93	0.9992
Glucosamine	2.53	0.9995

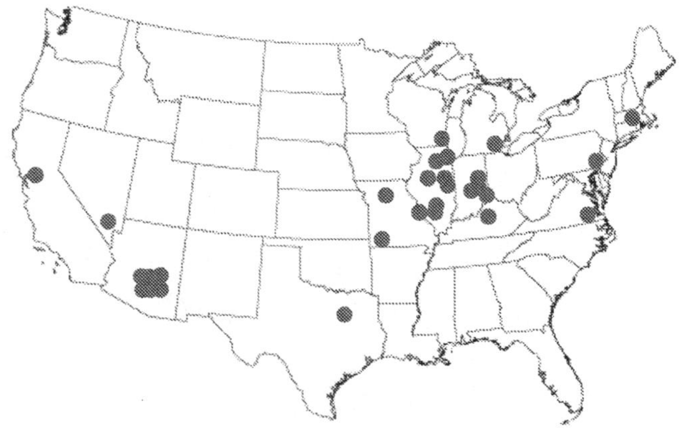

Figure 2.1 Geographic distribution of water treatment plants included in the two rounds of utility sampling campaigns

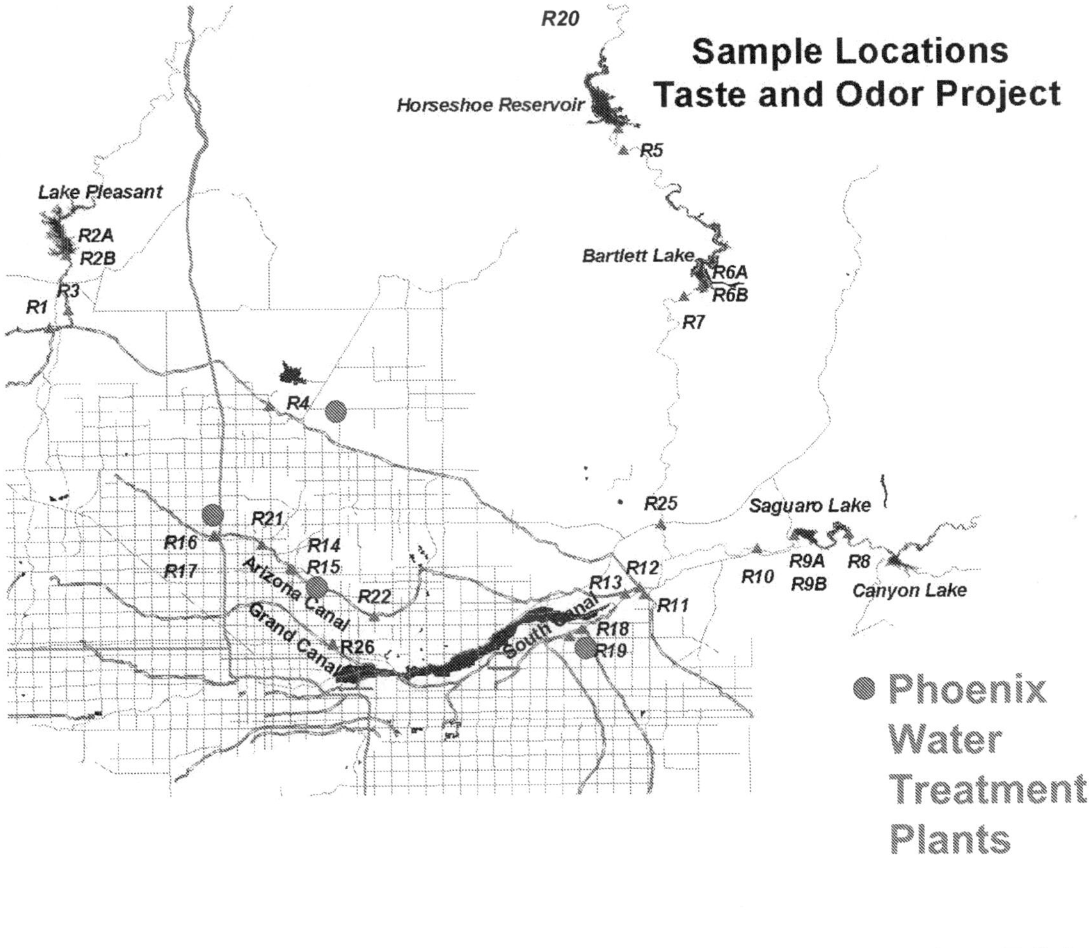

Figure 2.2 Map showing locations of reservoir and other sampling locations (R#) in Central Arizona

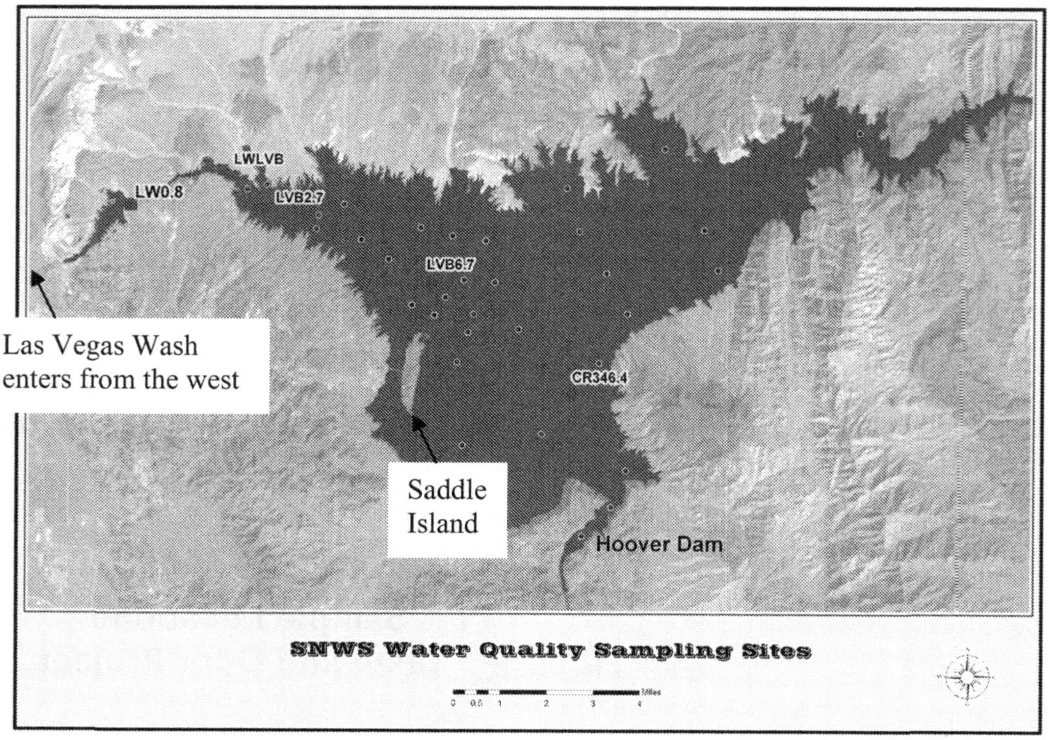

Figure 2.3 Sampling locations on Lake Mead, Nevada

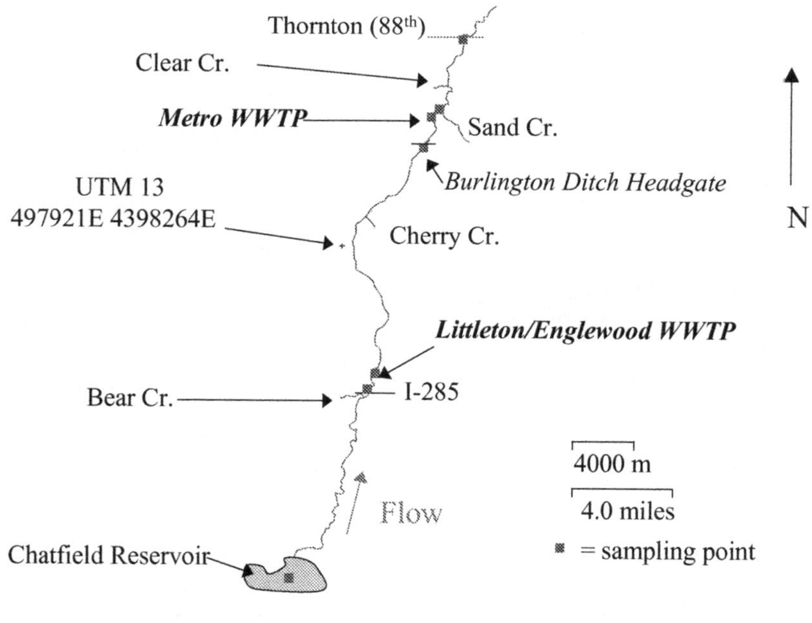

Figure 2.4 Sampling points along the South Platte River

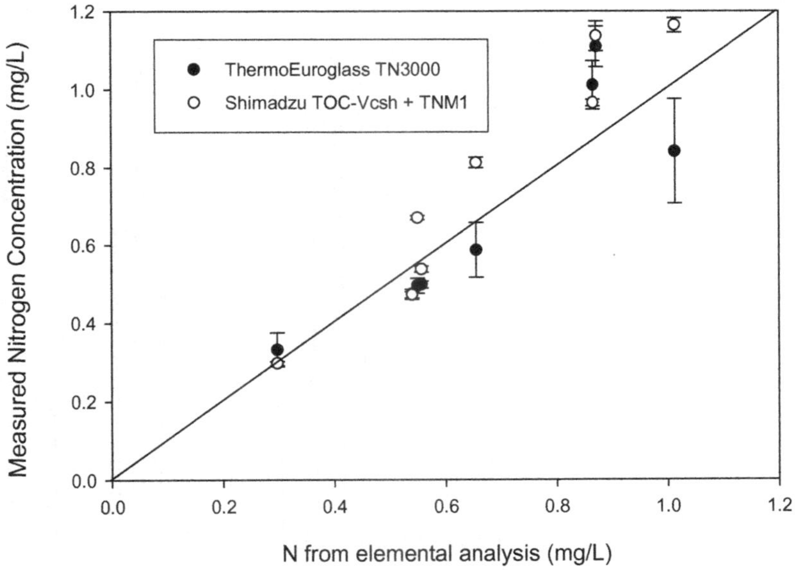

Figure 2.5 Nitrogen concentrations determined using Shimadzu TN and Euroglass TN3000 analyzers versus nitrogen estimated from elemental analysis

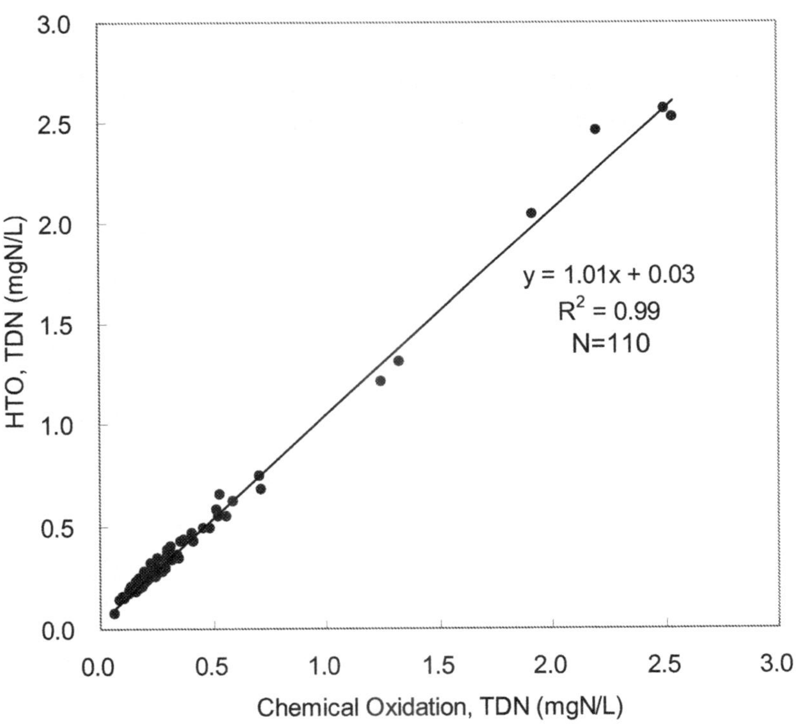

Figure 2.6 Comparison of TDN measurements by HTO and WCO methods on split samples.

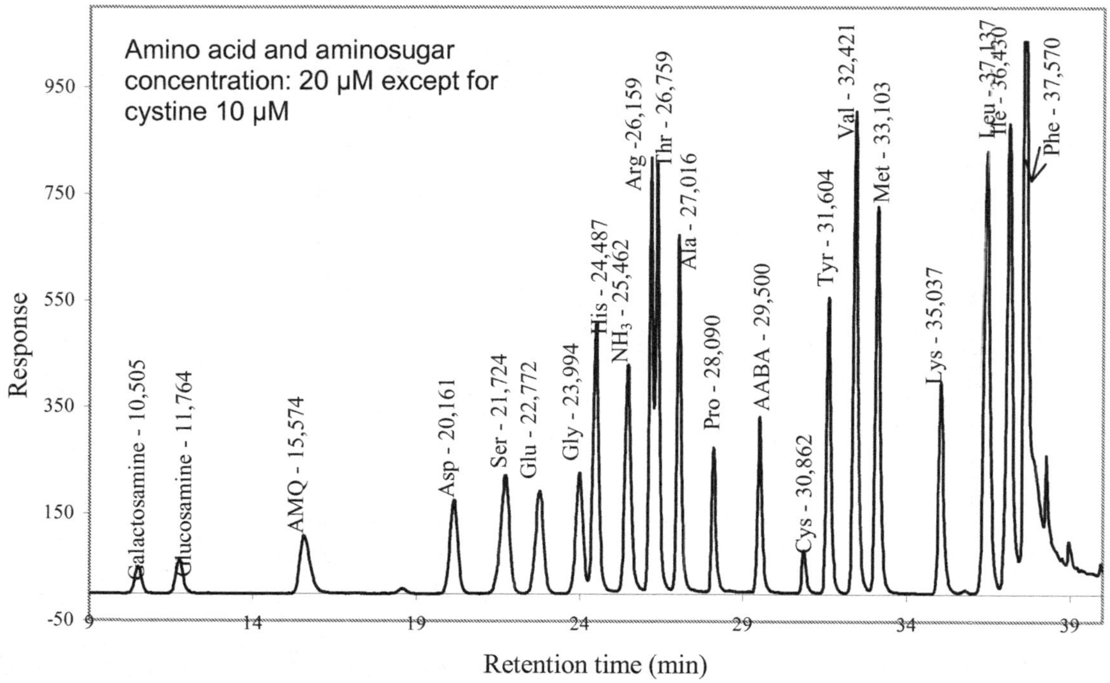

Figure 2.7 HPLC chromatogram for a standard solution containing 17 free amino acids and two aminosugars obtained after AQC derivatization.

CHAPTER 3 DEVELOPMENT OF A DISSOLVED ORGANIC NITROGEN METHOD FOR DRINKING WATERS

This chapter describes results from three different approaches to develop a direct DON measurement method or at least methods that increase the accuracy of DON measurement. All three approaches build upon the current approach for DON quantification that involves subtracting DIN concentrations from the TDN concentration:

$$DON = TDN - NO_3^- - NO_2^- - NH_3/NH_4^+ \qquad (3.1)$$

TDN measurement commonly involves preparative steps that convert all the nitrogen to inorganic nitrogen species (NO_2^-, NO_3^-, NH_3/NH_4^+, and/or NO_x gas) prior to quantification in aqueous or gaseous phases. The common wet chemical oxidation method, Total Kjeldahl Nitrogen (TKN, the sum of DON and ammonia nitrogen) method uses hot sulfuric acid to convert DON to NH_4^+. However, the TKN method generally has low precision (method detection limit (MDL) of 0.1 to 0.2 mgN/L) and suffers from incomplete oxidation of some nitrogenous compounds (APHA et al. 1998; Fleck and Munro 1965; Jones and Daughton 1985). Other wet chemical oxidation (e.g., persulfate oxidation) and high temperature combustion methods are also used to measure TDN and have better precision (Ammann et al. 2000; Sharp et al. 2002; Westerhoff and Mash 2002). Several techniques measure DIN species, but colorimetric and ion chromatographic methods are the most common (Westerhoff et al. 2002). Each of these measurements has an acceptable variance of typically 3% to 5%. Thus during calculation of DON (equation 3.1), subtraction of multiple, independently measured concentrations propagates analytical variance in the calculated DON concentration (Devore 1995):

$$V(DON) = V(TDN) + V(NO_3^-) + V(NO_2^-) + V(NH_4^+) \qquad (3.2)$$

where V(X) is a variance of each parameter "X". Using a V(X) of 5% for each independent measurement and assuming a surface water has TDN = 2.0 mgN/L, NO_3^- = 1.8 mgN/L, and NH_4^+ = 0.05 mgN/L (selected values represent average concentrations from 30 surface water samples collected from 14 U.S. states; see Chapter 4), then the calculated DON concentration would be 0.15 mgN/L (Equation 3.1). However, V(DON) would be 0.19 mgN/L (Equation 3.2). Thus, V(DON) can be greater than the DON concentration, and a need exists to improve the accuracy of DON quantification. Therefore a direct DON method should remove DIN species with minimal loss of organic nitrogen, or an improved DON method should increase the accuracy of DON by sufficiently lower the DIN concentration. To this end, the following three approaches were investigated to develop a DON method suitable for analysis of drinking waters:

1. Catalytic reduction of nitrate and nitrite to nitrogen gas or ammonia
2. Adsorption of ammonium ion onto zeolite
3. Permeation of nitrate, nitrite, and ammonia across dialysis membranes

Representative results of studies evaluating the first two approaches are presented below. However, these approaches only showed marginal success. In contrast the third approach of using dialysis as a pretreatment to remove DIN species was quite successful, and most of this chapter is dedicated to demonstrating the development of this method. As a result of the improved accuracy in DON measurement with use of the dialysis DIN pretreatment method, this DIN pretreatment step is the basis of most of the results presented in subsequent chapters of this report.

Other methods for removing nitrite, nitrate, and ammonia were considered and some preliminary experiments conducted. Ion exchange resins remove nitrate and nitrite, but also remove organic acids from solution and potentially leach organic nitrogen from the ion exchange resins. Ammonia stripping at elevated pH was evaluated, but the process exhibited fairly slow kinetics and it was difficult to get >90% removal as the driving force for gas-liquid stripping decreased as the liquid phase ammonia concentration was lowered. Partitioning of organic matter onto solid-phase extraction media (e.g., 500-mg hydrophilic-lipophilic balance (HLB) cartridges from Waters Corp. (Millford, MA)) was evaluated as an approach to separating DON from DIN. However, DON adsorption was not 100% and the calculation of DON still required subtraction of multiple nitrogen measurements. Extraction of DON into a solvent phase and then direct quantification of DON was considered, but not explicitly studied since solvent injection into the Shimadzu TOC-V instrument was not appropriate. However, addition of research along these lines may be fruitful for the development of a direct DON measurement if high DON recoveries can be obtained. The difficulty resides in the fact that DON is present in different NOM fractions, each with unique polarity (acids, bases, neutrals) and variable abilities to be removed from solid-phase resins (e.g., colloid fraction). Therefore, the three methods described appeared to be the most promising to improve DON measurement in drinking water and reclaimed wastewater samples.

CATALYTIC REDUCTION OF NITRATE AND NITRITE

The technique that was tested is a catalytic reduction of nitrate and nitrite into nitrogen gas. This technique has been developed at the Laboratoire de Catalyse en Chimie Organique (LACCO) of the University of Poitiers (France) to obtain selective reduction of nitrate into nitrogen, without the undesired formation of ammonia. This technique was proposed as a treatment of waters containing high content of nitrate. Results obtained with various bimetallic catalysts showed that total reduction of nitrate and nitrite (99%) occurs within 20 minutes in presence of hydrogen. Experiments were conducted with ultra-pure water spiked with high concentration of nitrates (1 mM – 62 mg/L NO_3^-), no data were available on the effect of catalytic reduction on DON. The experiments conducted in this study consisted of catalytic reduction tests in absence and in presence of model compounds (amino acids) or DON from several isolates of surface waters. Several catalysts supplied by the LACCO and prepared in the laboratory were tested.

Bimetallic Catalytic Reduction Theory

In the presence of Pt-Cu/Al_2O_3 catalyst, the first step in the nitrate reduction is a redox reaction between Cu° and NO_3^-, leading to NO_2^- intermediates or directly to N_2 or NH_4^+, and to an oxidized form of copper. Subsequently, the role of the precious metal is to activate hydrogen, allowing the reduction of Cu^{2+} according to the following scheme for nitrate reduction:

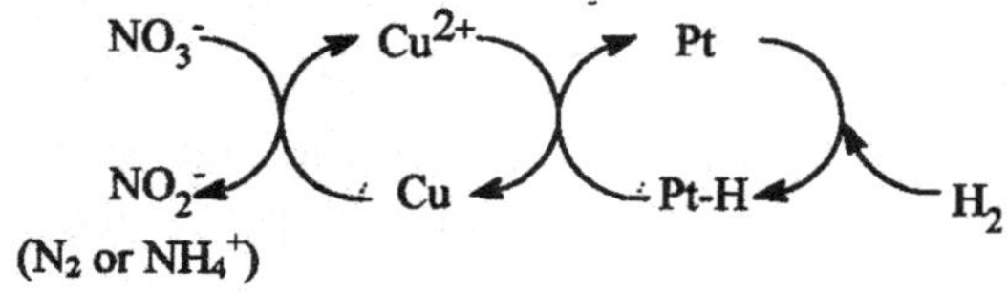

Copper also reduces nitrite to nitrogen and to a lesser extent ammonia that are the final by-products. Other bimetallic catalysts can be used such as Pd-Cu/Al_2O_3, Pt-Cu/Al_2O_3, Pd-Sn/Al_2O_3, Pd-In/Al_2O_3 and Pd/SnO_2, which differ in their activity and selectivity for nitrate reduction. Activity is quantified as the amount of nitrate reduced per minute and per g of catalyst. A kind of negative selectivity is given in terms of the formation of the undesired by-product ammonium, which means that the more ammonium is formed, the less selective is the catalyst. Depending on the amount (% wt.) of noble and promoting metals, activity and selectivity range between 2 and 7 x 10^{-5} mol/min-$g_{catalyst}$ and between 96 to 99 % formation of N_2 for Cu-Pd and Cu-Pt catalysts, respectively. The methods of preparation are also of major importance on the activity of catalyst. Preparation method by catalytic reduction has been shown to give better results than by successive impregnation by cationic exchange. Recent results obtained by the LACCO showed that the Pd-Sn/Al_2O_3 and Pd/SnO_2 bimetallic catalysts have the highest activity and selectivity for nitrate reduction.

Preparation of Supported Metal Catalysts

Different types of catalysts were used. In a preliminary set of experiments, three catalysts (Pd-Sn/Al_2O_3, Pd-In/Al_2O_3 and Pd-SnO_2) prepared by the LACCO were tested. They were prepared by wet impregnation using ammonium salt of Pd and nitrate salt of In and Sn. The composition of catalysts was as follows:

- Pd (5 wt.%) -In(1.75 wt.%)/Al_2O_3,
- Pd(5 wt.%)-Sn(1.75 wt.%)/Al_2O_3
- Pd(5 wt.%)/SnO_2

Two series of Pd(5 wt.%)-Sn(1.75 wt.%)/Al_2O_3 and Pd(5 wt.%)/SnO_2 catalysts were also prepared in our laboratory. Pd-Sn/Al_2O_3 catalysts were prepared by deposition-precipitation. The support material, γ-alumina (Procatalyse), was suspended in a 2% Na_2CO_3 solution for 15 minutes. A solution of Na_2PdCl_4 salt was slowly dropped into the solution under vigorous stirring and Pd was let to precipitate on alumina for 2 hours. The second metal was then introduced in the desired amount as a metal chloride solution in the same manner. The solution was kept under stirring over night and then the catalyst was reduced by adding a 20% sodium formate solution. A first series of the catalyst was used after simple drying at 80°C. The second series was further calcined under N_2/O_2 (60/20 mL/min) gas mixture at 400°C. Pd/SnO_2 catalysts were prepared by wet impregnation of Pd on tin oxide. SnO_2 was prepared from a NaOH-$SnCl_2$ solution (pH 11). Tin hydroxides were filtered, dried over night at 105°C and then calcined at 400°C under N_2/O_2 (80/60 mL/min) gas mixture. The desired amount of palladium chloride salt was dissolved in a small amount of Milli-Q water and pH was then increased to pH 11 with NaOH. The tin oxide was slowly added into the solution and the solution was kept under stirring over night. As for Pd-Sn/Al_2O_3 catalysts, a series of Pd/SnO_2 catalyst was used after drying at 80°C. The second series was calcined as described above.

Experiments were carried out in a 250-mL batch reactor vessel at room temperature. The catalyst (0.1 g) was previously introduced in the reactor and degassed under N_2 for 15 minutes to remove dissolved and adsorbed oxygen. Then 100 mL of the tested solution, previously purged with N_2, were also introduced in the reactor and hydrogen was bubbled through a porous glass. Catalyst dispersion was operated by the hydrogen bubbling. After a given reaction time, an aliquot of the solution was withdrawn and was analyzed for organic and inorganic nitrogen content.

Experiments on model organic compounds

To study the stability of organic nitrogen during catalytic reduction a solution prepared by diluting a 2.5 mmol/L mixed amino acids standard (Pierce) in ultra-pure water (18 ΩM Millipore system) was analyzed before and after 30 minutes of catalytic reduction. Three catalysts Pd-Sn/Al_2O_3, Pd-In/Al_2O_3 and Pd-SnO_2 were tested. Concentration of each amino-acid was 10 μmol/L except that the concentration of L-cystine was 5 μmol/L.

Figure 3.1 shows both chromatograms of the amino acids before and after catalytic reduction with Pd-In/Al_2O_3. Table 3.1 presents the values of total nitrogen analyzed before and after catalytic reduction. Results in Table 3.1 shows that the analysis of the initial concentration of total nitrogen gave 3.20 ± 0.32 mg/L, similar to the theoretical nitrogen content calculated from the amino acid formula (3.29 mg/L). Results showed also that except for Pd/SnO_2 the concentration of total nitrogen after catalytic reduction increased dramatically in agreement with the analysis of ammonia after catalytic reduction with Pd-In/Al_2O_3 (Figure 3.1). Also, since these catalysts were prepared from $Pd(NH_3)_4(NO_3)_2$, the release of both ammonia and nitrate from catalysts into the solution is possible and would explain the results obtained with Pd-In/Al_2O_3 and Pd-Sn/Al_2O_3. In response, new catalysts with metal chloride salts were prepared.

For Pd-In/Al_2O_3 catalyst results in Figure 3.1 show that the chromatograms differ in some extent after catalytic reduction. The first significant result is the large increase of the ammonia peak. The response of most amino acids remained unchanged after catalytic reduction, only few amino acids being affected. Methionine and cystine that contain C-S and disulfur bonds, respectively and histidine that has double C-C and N-C bonds had their response to decrease. Only one significant increase was observed for alanine.

Figure 3.2 shows the effect of reductive catalysis on amino acids for the three palladium-based bimetallic catalysts. Results are expressed as the ratio between the response area of amino acid after and before reductive catalysis. Results show that ammonia response remained constant after catalytic reduction for both Pd/SnO_2 and Pd-Sn/Al_2O_3 and that for Pd/SnO_2 methionine and histidine were stable during catalytic reduction. Cystine was reduced by the three catalysts. The increase of alanine concentration was only observed for Pd-Sn/Al_2O_3 and Pd-In/Al_2O_3.

Similar results (reduction of sulfurous amino acids and histidine) were also observed for the Pt-Cu/Al_2O_3 catalyst (results not shown). This catalyst was not kept for further tests since recent results obtained by the LACCO for the reduction of nitrate have shown that this catalyst exhibits a lower activity and selectivity than Pd-Sn/Al_2O_3 and Pd/SnO_2 catalysts.

The decay of some amino acids can be explained by their specific structure. The decrease of methionine and cystine is probably due to the hydrogenation of the C-S and S-S bonds. Hydrogenation of cystine is known to produce cysteine. Further hydrogenation of the thiol group of cysteine may lead to the formation of alanine, which would explain the observed increase of this amino acid. According to the chromatograms, 50 % of the cysteine would be reduced into alanine for Pd-In/Al_2O_3 and Pd-Sn/Al_2O_3. In case of Pd/SnO_2, the hydrogenation of cystine would stop with the formation of cysteine because the formation of alanine was not observed.

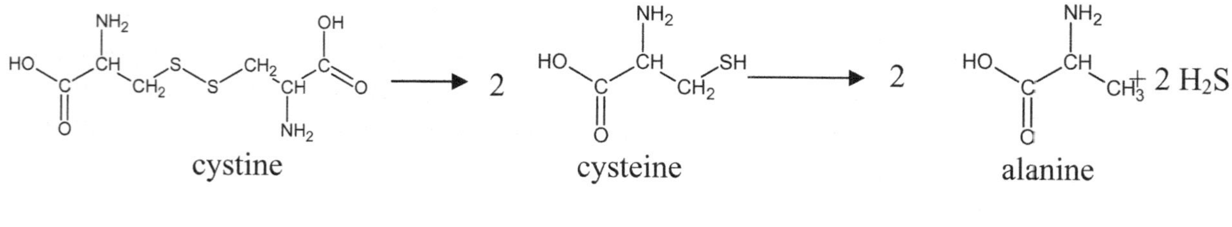

The first hypothesis for the disappearance of methionine is the hydrogenation of the C-S bond leading to the formation of methanethiol and 2-aminobutyric acid (mechanism 1). Alanine is probably not formed from methionine because the reaction would correspond to the cleavage of the C-C bond. However, the formation of 2-aminobutyric acid, which is also the internal standard was not observed in our experiments. Then, the most probable mechanism is the formation of 2-amino-4-mercaptobutyric acid and methane (mechanism 2).

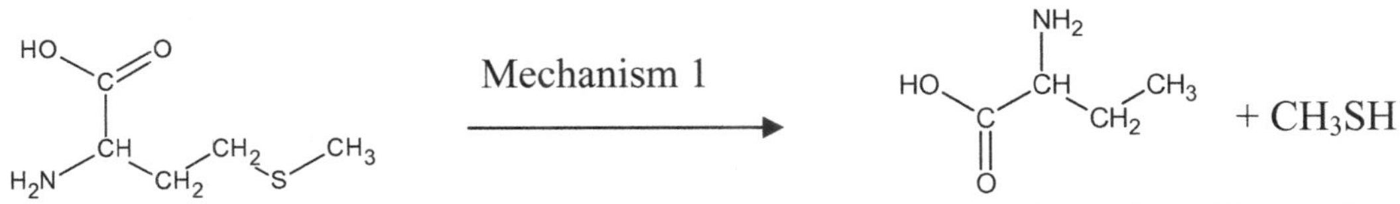

methionine 2-aminobutyric acid methanethiol

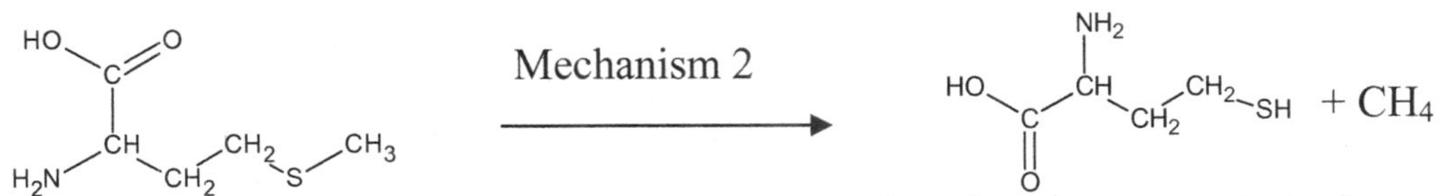

2-amino-4-mercaptobutyric

Hydrogenation of histidine would occur on the C-N and C-C double bonds as follows:

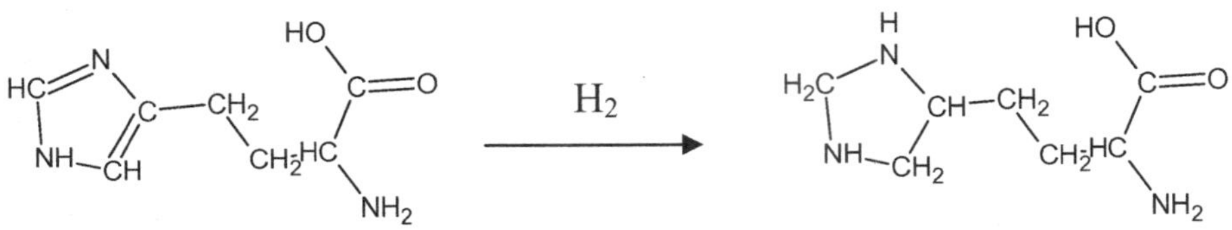

From these results, it was concluded that the amine group is not reduced into ammonia since no formation of ammonia was observed for the two catalysts and that the presence of ammonia for the other catalysts was attributed to ammonia coming from the catalyst. Dissolved organic nitrogen from amino acid is then probably stable during catalytic reduction in our conditions. New catalysts were prepared with chloride salts instead of ammonia salts to prevent contamination of the aqueous solution and were tested for catalytic reduction of nitrate.

Catalytic Reduction Tests of Nitrate

Catalytic reduction in absence of NOM

In a first run of experiments, the first series of catalysts prepared without the calcination step were tested for the catalytic reduction of nitrate in absence of NOM. A kinetic experiment was performed with Pd-Sn/Al_2O_3 for an initial concentration of 50 mg/L NO_3^- (11.3 mg/L N-NO_3^-). The solution was buffered during the experiment by bubbling H_2/CO_2 into the reactor. Nitrate and total nitrogen were analyzed at different time intervals.

Figure 3.3 shows the evolution of both nitrate and total nitrogen for 3 hours of catalytic reduction. Results show that nitrates are rapidly reduced within the first 20 minutes. After 30 minutes, the residual concentration of nitrates was below the detection limit (10 µg/L NO_3^-). Ionic chromatography analysis of both nitrate and nitrite indicated that nitrite was not formed during catalytic reduction. Analysis of total nitrogen shows also a fast decrease. After 15 minutes of reaction time, a residual of 2 mg N/L remained in solution. Specific analysis indicated that the residual concentration of nitrogen was due to the presence of ammonia.

Initial activity of the catalyst was estimated to be 8.5 mg N-NO_3^-. g^{-1} cata. h^{-1} (0.6 mM N-NO_3^-. g^{-1} cata. min^{-1}). The negative selectivity expressed in terms of ammonium formation was about 19% (as nitrogen), value similar to the catalysts prepared from ammonium salt of pd..

Additional experiments were also performed with both Pd/SnO_2 and Pd-Sn/Al_2O_3 catalysts, with concentrations of nitrates equal to 2.26 and 11.3 mg N-NO_3^-/L (10 and 50 mg/L NO_3^-) and with solution buffered with phosphate (KH_2PO_4 10 mM) or carbonate (continuous bubbling of CO_2). Table 3.2 shows the concentrations of nitrates and total nitrogen before and after 3 hours of reduction for all conditions studied. Results show that both catalysts exerted a strong reduction capacity of nitrate ion. Except for experiment 1 with Pd-Sn/Al_2O_3, the concentrations of nitrate were below the detection limit of 10 μg N-NO_3^-/L. For the two buffers, the final pH ranged between 4 and 6, which allows a good activity of the catalysts. The concentrations of total nitrogen after 3 hours of reduction ranged from 1.40 mg/L and 2.0 mg/L for the initial concentrations of 2.26 or 11.30 mg N-NO_3^-/L. For experiments 2 to 5, ammonia did represent 100% of the total nitrogen content after 3 hours of reduction. The percentage of ammonia formed during the catalytic reduction was about 66% and 20% for the initial concentration of nitrates of 10 and 50 mg/L, respectively. Thus, the selectivity decreased when the concentration of nitrate decreased in agreement with literature data. This is explained by the fact that formation of 1 mole of N_2 needs 2 moles of nitrate whereas the formation of NH_4^+ only needs one mole of nitrate. Then, the formation of N_2 is favored for high concentration of nitrate adsorbed on the catalyst.

Catalytic Reduction in Presence of NOM

In a first experiment, the effect of catalytic reduction on DON was tested with a NOM isolate (HPO fraction, N =1.37%) extracted from the Vienne River (France). Total nitrogen was analyzed before and after a 3-hour catalytic reduction using Pd-Sn/Al_2O_3 and Pd-SnO_2 catalysts. Both catalysts were not calcined. Results are given in Table 3.3. Results show a decrease of 10 and 19% (sorption, degradation) of the DON content after catalytic reduction, which is consistent with previous results obtained with amino acids (Figure 3.4).

A synthetic water was prepared from sodium nitrate and the same HPO fraction from Vienne River (France). The concentrations of nitrates and organic nitrogen were 11.3 mg/L N-NO_3^- and 0.56 mg/L, respectively. The DOC of the solution was 20 mg/L. Catalytic reduction of an aliquot of 100 mL was conducted with Pd-Sn/Al_2O_3 catalyst. Samples were taken at different time intervals and were analyzed for nitrates and total nitrogen. Figure 3.4 shows the evolution of nitrates and total non-purgable nitrogen versus reaction time. For comparison, results obtained in absence of NOM were also plotted on the same figure. Results show that nitrates are reduced within 1 hour in the presence of NOM whereas in absence of NOM nitrates were reduced within the first 20 minutes. Also, in presence of NOM, total nitrogen was only slightly reduced from 9.3 to 8.2 mg/L. Since both nitrates and nitrites were not detected, total nitrogen was considered to be the sum of ammonia and organic nitrogen (≈ 0.56 mg/L). As a consequence, the selectivity expressed in terms of N_2 production was calculated to be equal to only 10% in the presence of NOM.

To validate this first result in presence of NOM, a similar experiment was conducted with Pd-Sn/Al_2O_3 catalyst prepared with palladium and ammonium salts and previously washed with ultrapure water. The same isolate of NOM with the same DON concentration (0.56 mg/L DON) was used for comparison with previous results. Ammonium ion was analyzed by ion chromatography. Figure 3.5 shows the evolution of nitrates, total nitrogen and ammonia versus reaction time. Results show that nitrates were totally reduced within more than 1 hour. Ammonia was not present at the initial time indicating that the washing procedure was efficient to remove ammonium traces from catalyst. Ammonia was quantitatively formed from nitrate ion (100% formation of ammonia). At the end of the catalytic reduction, the sum of analyzed ammonia (10.4 mg N/L) and introduced DON (0.56 mg N/L) is approximately similar to the initial total nitrogen content analyzed (10.8 mg N/L).

Similar results were obtained with different NOM isolates. Figure 3.6 summarizes the evolution of $N\text{-}NH_4^+$ and $N\text{-}NO_3^-$ during catalytic reduction in absence and in presence of NOM. These results confirmed that the presence of NOM slowed the kinetics of nitrate reduction and led to higher percentage conversions into ammonia i.e. both activity and selectivity of bimetallic palladium based-catalyst decrease in presence of NOM. This observation can be explained by the sorption of NOM on active sites of the catalyst.

Also, when catalysts were added to buffered pure water, conductivity measurements showed that metals leached from the catalysts. Several washings with pure water partially reduced the phenomenon but were not sufficient to limit the clogging of the injection needle of the Euroglass nitrogen analyzer TN3000 after several injections. This clogging was probably due to the formation of metal oxide in the oven (900 – 1000 °C) and was not possible to remove by physico-chemical treatments such as sonication or acid treatment. Therefore, two new catalysts $Pd\text{-}Sn/Al_2O_3$ and Pd/SnO_2 were synthesized by including a calcination step performed under N_2/O_2 (60/20 mL/min) gas mixture at 400°C. This step facilitated the deposit of tin and palladium chloride and improved the stability of the catalysts.

Preliminary experiments showed that these new catalysts reduced nitrate in ultra pure water with similar activity and selectivity than precedent catalysts. Further tests were then performed to study the stability and adsorption of organic nitrogen on these two catalysts. Experiments were conducted with reconstituted waters prepared by diluting about 20 mg of NOM isolates in 500mL ultra pure water. Six NOM isolates extracted from rivers and reservoir were studied. The concentrations of organic carbon and organic nitrogen, the UV absorbance and pH of NOM solutions are given in Table 3.4. Experimental procedure was as follows:

- N_2 bubbling removed dissolved oxygen adsorbed on the catalyst (0.1 g),
- H_2 bubbling (125 mL/min) allowed pre-reduction of palladium to ensure that only Pd° species was present on catalyst,
- H_2 was purged by N_2/CO_2 (1:1 125 mL/min) bubbling and 100 mL NOM solution, previously degassed by N_2, was introduced into the reactor. After 3 hours of contact time, an aliquot of the solution was withdrawn and filtered prior analysis to control NOM adsorption on catalyst,
- N_2/CO_2 was then replaced by H_2/CO_2 (1:1 125 mL/min) for 3 hours. After 3 hours of reduction, an aliquot of solution was also withdrawn and filtered prior analysis to control the stability of organic nitrogen under catalytic reduction.

Figures 3.7 and 3.8 show the removal of organic carbon, organic nitrogen and UV absorbance after the first three hours of contact time with $Pd\text{-}Sn/Al_2O_3$ and $Pd\text{-}SnO_2$ catalysts, respectively. Isolate numbers correspond to those in Table 3.4. Results showed that the removal of organic carbon and organic nitrogen ranged between 65 to 78% and between 23 to 39% for $Pd\text{-}Sn/Al_2O_3$ and $Pd\text{-}SnO_2$ catalysts, respectively. Removal of UV absorbance was even higher with up to 98 % for Pd-Sn/Al_2O_3 catalyst and isolate 2 (Seine River TPI). Adsorption of organic carbon depended mainly of the nature of the catalyst and was approximately 15 (± 2) and 6 (± 1) mg C/g catalyst for Pd-Sn/Al_2O_3 and Pd-SnO_2, respectively. For organic nitrogen, adsorption ranged between 0.2 mg N/g catalyst for Pd/SnO_2 with isolate 6 (Gartempe River TPH) to 1.9 mg N/g catalyst for Pd-Sn/Al_2O_3 with isolate 1 (Ribou Reservoir TPH). Similar UV abs., organic carbon and organic nitrogen residuals were analyzed after 3 more hours of contact time with H_2/CO_2 bubbling. Also, adsorption tests performed at pH 11 with the same NOM solutions showed identical adsorption.

To conclude both catalysts stabilized by calcination showed high NOM adsorption capacity. In gaseous phase catalysis, palladium-based catalysts are known to be rapidly contaminated by organic carbon. In our study, calcination step probably increased the Pd concentration on catalyst surface, which may then explained the higher adsorption removal observed for the new catalysts. Considering these results, the reduction of nitrate in presence of NOM was not conducted with new catalysts because the lost of organic nitrogen by adsorption is incompatible with its analysis. Further research is needed to develop new catalysts that would not interact with NOM (i.e. that would not adsorb NOM and would have similar conversion yield of nitrate into N_2 in presence and in absence of NOM).

ADSORPTION OF AMMONIA ONTO ZEOLITE

Ammonia adsorption onto zeolites has been well studied (Jorgensen and Weatherley 2003; Jung et al. 2004). Removing ammonia could improve DON determination in at least two different circumstances. First, TKN represents ammonia plus organic nitrogen measurement. Ammonia removal would allow TKN methods to quantify just organic nitrogen, if the ammonia could be selectively removed. Second, conversion of nitrate and nitrite to ammonia via catalytic reduction would subsequently require ammonia addition in order to quantify organic nitrogen. The key question investigated regarding the use of zeolites centered around assessing organic matter (i.e., DON) adsorption onto the zeolites under conditions capable of removing nearly 100% of the ammonia from solution.

Batch adsorption experiments were conducted using a commercially available zeolite material (CABSORB ZS403H; GSA Resources, Inc., AZ). Experiments were conducted in three different water matrices (Nanopure water, tap water and a model water prepared from a NOM isolate), each spiked with ammonia. Variable dosages of zeolite was added to the liquid samples, such that the ratio of zeolite volume to water volume varied from 0.01% to ~ 12%. Both ammonia and TDN concentrations were measured in triplicate experiments. Zeolite addition resulted in up to 97% ammonia removal in deionized water and 91% ammonia removal in tap water at very high zeolite dosages (Figures 3.9 and Table 3.4). Complete ammonia removal, to levels below the MDL for ammonia, was not achieved, even at zeolite dosages higher than those presented below. Higher zeolite dosages were required in the tap water than Nanopure water experiments for comparable ammonia removals, presumably due to adsorption of other aqueous cations (calcium, magnesium, sodium, potassium) present in the tap water (TDS ~ 600 mg/L). However, it appeared that zeolite addition could effectively reduce ammonia concentrations using approximately a 5% volume ratio of zeolite to water.

Separate experiments were conducted with a different zeolite powers (Zeolite 5A, LTA type, exchanged with Ca^{2+}). Kinetic tests conducted with 3 mgN-NH_4^+/L and 9 g/L zeolite achieved 94% ammonia removal within 60 minutes.

Addition experiments were undertaken to evaluate organic matter removal upon zeolite addition. One set of results are presented in Table 3.1, and represent removals of a hydrophobic acid fraction of NOM (isolated from Saguaro Lake, Arizona). At zeolite dosages (v/v) of 2% and 5% approximately 5% to 10% of the organic matter, including organic nitrogen, was adsorbed by the zeolite. Other NOM fractions (e.g., bases or colloids) would be expected to adsorb onto zeolite as well. Overall, while zeolites appear effective at reducing ammonium ion concentrations, some removal of organic matter will always occur. In order to achieve very high ammonia removal (e.g., low residual ammonia) using zeolites to facilitate more accurate DON determination (e.g, Equation 3.1 and 3.2), unfortunately some DON would be removed by the zeolite as well. Thus, the use of zeolites was not considered further in this study.

DIALYSIS PRETREATMENT FOR DIN/DON SEPARATION

The concept of dialysis pretreatment is that physical separation of DIN from DON occurs, improving DON determination (Equation 3.1 and 3.2). It was hypothesized that DIN would permeate the dialysis system, but DON would be retained inside the dialysis system. Measurement of nitrogen species retained inside the dialysis system would have a lower DIN/TDN ratio (higher DON/TDN ratio) than the initial sample due to DIN permeation across the dialysis system. The separation concept and schematic dialysis apparatus is shown in Figure 3.10 (photograph in Figure 3.11).

The dialysis pretreatment method development included batch experiments to evaluate effects of sample pH, sample conductance, temperature, organic matter biodegradation, and adsorption of NOM isolates onto the dialysis membrane. Furthermore, continuous flow dialysis experiments using the apparatus depicted in Figure 3.10 was conducted with inorganic nitrogen and model organic nitrogen compounds in Nanopure water, surface water, tap water, treated wastewaters, and model waters containing NOM isolates. Details for developing the dialysis pretreatment method are described elsewhere (Lee and Westerhoff Accepted). In all cases, the dialysis pretreatment approach effectively separated DIN from DON, and thus was considered a major advance. As such dialysis pretreatment was used in subsequent most of the work presented in later chapters of this report. An overview of the method and step-by-step procedures are described below. The experimental protocols developed by ASU were successfully implement at Univ. of Poitiers.

Dialysis Pretreatment System

Dialysis pretreatment used 20 mm diameter cellulose ester (CE) dialysis tubes (Spectra/Por, Spectrum Laboratories Inc., CA). The membrane is rigid and symmetric with a nominal MWCO of 100 Da. All membranes were cleaned prior to use by rinsing thoroughly with deionized water over several days until DOC was less than 0.01 mgC/L in the supernatant solution, indicating minimal leaching from the membrane. Dialysis tubes were clamped at the bottom, filled with 25 mL of sample, and clamped at the top. The ratio of membrane surface area to sample volume was maintained at 2.0 cm^2/cm^3. The clamped dialysis tubes were suspended in custom-made plexiglass supports in a Nalgene container, which was wrapped with foil to minimize sample exposure to light. Inlet and outlet fittings were placed into opposite ends of the Nalgene container to allow the flow of a dialysis acceptor solution (distilled water) through the dialysis system. The acceptor solution was controlled by a flow meter, and passed just once through the dialysis system. The flowrate was set at 100 mL/min, and an overall ratio of sample volume (9 samples at 25 mL per sample) to cumulative acceptor volume (100 mL/min for 24 hours) was maintained at 1:640. A magnetic stir bar placed in the bottom of the Nalgene tank facilitated mixing. Subsequent to this study the Investigators have made additional continuous flow reactors capable of holding up to 24 dialysis bags, but a constant sample volume to cumulative acceptor volume has been maintained at 1:640.

Effect of Dialysis Duration

The kinetics of DIN permeation across the dialysis membrane and effect of dialysis time on potential NOM loss from the sample were evaluated. Figure 3.12 presents the concentration of DIN species retained within the dialysis tubes from samples removed from the dialysis system and sacrificed over a period of five days. When starting with the same initial concentrations, nitrate and ammonia permeation kinetics are equivalent. Ammonia and nitrate concentrations were reduced by 70% to 80% in 24 hours and eliminated (below the MDL) in 120 hours. Separate experiments indicated that the percentage removed over a specified time is dependent upon the initial DIN specie concentration.

Figure 3.13 presents the change in DON as a function of dialysis time. For the model compounds (except urea) and other samples there no difference in DON concentration between the initial non-dialyzed sample and one dialyzed for 24 hour. After 120 h of dialysis the concentrations of tryptophan (MW = 204 g/mol) and tripeptide (MW = 307 g/mol) did not change in the same time ($\alpha = 0.05$). However 97% of the urea (MW = 60 g/mol) was removed. There was no difference ($\alpha = 0.05$) between the initial concentration and the concentrations over time of a hydrophobic NOM isolate. The Arizona Canal water sample also yielded a good DON recovery, but a slight loss of DON (~5%) occurred from 72 hours. The Mesa WWTP sample lost 10% to 15% of the its DOC during prolonged dialysis. Parkin and McCarty reported that less than 10% of the DON in a WWTP (activated sludge process) effluent consisted of free or combined amino acids and that the activated sludge effluent is about $6 \pm 9\% < 165$ Da fractions (Parkin and McCarty 1981). Therefore, some of the DOC lost could have been free low molecular weight amino acid materials. Overall, organics with molecular weights of less than nominal molecular weight cutoff of the dialysis tubes (100 Daltons) may permeate across the dialysis membrane. However, most organic materials in natural water generally have molecular weights larger than 100 Da (Taylor et al. 1987).

From the above tests, loss of DON or DOC during dialysis could be attributable to either permeation across the dialysis membrane or adsorption onto the dialysis membrane. Batch adsorption tests were conducted with square pieces of dialysis tube material placed into glass bottles with water samples. Adsorption (not dialysis) isotherm experiments were conducted with these cut-pieces of dialysis membranes. DIN did not adsorb onto the membrane (not shown). Other results are illustrated in Figure 3.14 for model compounds and NOM isolates and Figure 3.15 for surface waters and treated wastewaters. While some amino acids showed no adsorption (tryptophan, glutamic acid, arginine, and glycine), at high dialysis membrane surface area aniline ($\log K_{ow} = 0.90$) and tributylamine ($\log K_{ow} = 4.6$) had up to 25% partitioned to the membrane. The hydrophobic NOM fraction showed no adsorption, but a colloidal NOM fraction adsorbed, up to 15% as the membrane surface area increased. Neither DOC nor DON adsorbed onto the dialysis membrane from Arizona Canal water. However, in the treated wastewater samples (NDN WWTP effluent and TF WWTP effluent), up to 12% of the DOC and 8% of the TDN were adsorbed in the presence of the dialysis membrane (2 through 6 cm^2/cm^3). Cellular fragments (acetylamides) present in the colloidal fraction and wastewater samples could be adsorbed. Overall, less than 5% of DON adsorbed onto 50 cm^2 of dialysis membrane (membrane surface area to sample volume ratio = 2 cm^2/cm^3) in most cases.

Quantification of DON (Equation 3.1) without dialysis and after various durations of dialysis was investigated at different DIN/TDN ratios by spiking nitrate and ammonia into a local, filtered surface water. The ratio of nitrate to ammonia was constant in all experiments and equal to the ratio in the original surface water. The initial DON concentration in all samples were equal. Relative standard deviations in the TDN, nitrate and ammonia concentrations remained relatively constant (~3%) and independent of initial TDN concentration. DON was calculated by difference (TDN minus DIN; Equation 3.1); the relative standard deviation for the calculated DON values was 8%. The results are presented in Figure 3.16. The amount of DON in each test sample should be identical. However, as DIN/TDN ratios increased, the calculated mean DON slightly decreased, and the standard deviation of calculated DON concentration became larger (Equation 3.2). This result suggests that samples with high DIN/TDN ratios should be subjected to the dialysis pretreatment to reduce DIN levels and facilitate an accurate DON measurement. Dialysis of two samples with DIN/TDN ratios of 0.41 and 0.71 mgN/mgN were not different (α = 0.05) over the dialysis durations (24 hours). Samples with DIN/TDN ratios of 0.94 and 0.98 required 24 hours of dialysis before their calculated DON measurements were within the 95% confidence interval of the assumed "actual" DON concentration. These results indicate that the dialysis pretreatment leads to an accurate DON measurement in samples with high DIN/TDN ratio.

Dialysis Pretreatment Protocol

The final selection of dialysis conditions were optimized to achieve the following criteria:

- Achieve a high percentage of DIN permeation such that DIN species after dialysis contains < 50% of the TDN (i.e., DON accounts for the majority of TDN).
- Use sample volume to membrane areas ≤ 2 cm^2/cm^3 to minimize organic matter adsorption onto dialysis membrane surface.
- Minimize the total loss of DON or DOC during dialysis.

Based upon these three criteria a dialysis time of 24 hours was found to be the most appropriate and used for all DON measurements reported elsewhere in this report, unless explicitly stated otherwise. A final dialysis protocol (Table 3.6) was developed which was found suitable for drinking water samples and reclaimed wastewater samples with DOC concentrations less than 6 mg/L.

Comparison of DON measurements with and without the dialysis pretreatment

DON concentrations of bulk water samples (30 surface water and 30 finished drinking water) were compared with and without the dialysis pretreatment. Table 3.7 lists DON concentrations without dialysis pretreatment;11 of 60 samples had negative DON values due to cumulative analytical variability (Equation 3.2). DON measurements with the dialysis pretreatment were in the ranges of 0.06-0.43 mgN/L and 0.04-0.36 mgN/L for surface water samples and finished drinking water samples, respectively.

The distribution of differences between DON concentrations with and without the dialysis pretreatment was even (both positive and negative differences) (Figure 3.10). Most surface water samples and finished drinking water samples with a DIN/TDN ratio less than around 0.6 mgN/mgN showed no difference (α = 0.05). This implies that the dialysis pretreatment is not necessary for those samples. This result verifies a good DON recovery (no adsorption of organics onto the dialysis membrane and no loss of DON) in natural water during the dialysis pretreatment.

Figure 3.17 indicates that samples with a DIN/TDN ratio greater than 0.6 mgN/mgN have high analytical variations, and the variation increases as the DIN/TDN ratio becomes greater. More error is inherently possible with high DIN/TDN ratio water (i.e., equation 3.2). DOC with and without the dialysis pretreatment showed less than 10% difference in all the samples, and the average recovery of DOC was 98% (data not shown).

SUMMARY

A direct DON method would remove dissolved inorganic nitrogen (DIN) species with minimal loss of organic nitrogen, or an improved DON method would increase the accuracy of DON by sufficiently lower the DIN concentration. In drinking waters nitrate is the dominant DIN species Three approaches were evaluated to improve the accuracy of DON determination, which would be suitable for drinking water and reclaimed wastewater:

1. Catalytic reduction of nitrate and nitrite to nitrogen gas or ammonia
2. Adsorption of ammonium ion onto zeolite (e.g., after catalytic reduction of nitrate)
3. Permeation of nitrate, nitrite, and ammonia across dialysis membranes

The following conclusions were drawn from this work:

- Catalytic reduction of nitrate to gaseous nitrogen forms, other than ammonia, occurred in Nanopure water, but in the presence of NOM nitrate reduction produced mostly ammonia (and ammonium ion). Stripping ammonia gas from solution by elevating the sample pH was very slow kinetically. Therefore, catalytic reduction would not be suitable for reducing nitrate removal in natural waters, which would have improved the accuracy of DON determination through lowering of the DIN/TDN ratio. Low DIN/TDN ratios equate to higher DON/TDN ratios, which mean that calculation of DON by difference between TDN and inorganic nitrogen species could be more accurate.
- Zeolite adsorption of ammonium ion was effective (>90%). However, it was not possible to remove 100% of the ammonium ion. To achieve high ammonium ion removal, high zeolite dosages were required and resulted in adsorption of NOM which likely contained organic nitrogen. Zeolites may have a place in ammonium ion removal for poorly treated wastewater (i.e., trickling filters).
- Dialysis pretreatment for 24 hours effectively reduced DIN concentrations, thus improving accuracy of DON calculations. For most of the drinking waters dialysis completely removed the DIN species to below detection levels. Dialysis pretreatment is necessary for samples with DIN/TDN ratios greater than approximately 0.6 mgN/mgN; otherwise direct calculation of DON (equation 3.1) is accurate.
- Based upon the improved accuracy and low level detection of DON using dialysis pretreatment, all subsequent DON concentrations for natural waters in this report were quantified using dialysis pretreatment with TDN determination by high temperature catalytic oxidation.

Table 3.1 Catalytic reduction effect on total nitrogen content of amino acid standard

Catalyst	Total nitrogen before catalytic reduction (mg N/L)	Total nitrogen after catalytic reduction (mg N/L)
Pd-In/Al_2O_3	3.20 ± 0.32	21.7 ± 0.34
Pd-Sn/Al_2O_3	3.20 ± 0.32	7.50 ± 0.36
Pd-SnO_2	3.20 ± 0.32	3.70 ± 0.44

Table 3.2 Effect of catalyst, buffer and concentration of nitrates

n°	Known N-NO_3^- initial * (mg/L)	N_T before reduction (mg/L)	Catalyst	Final pH	Buffer	N_T after reduction (mg/L)	N-NO_3^- after reduction
1	2.26	2.60	Pd-Sn/Al_2O_3	5.39	KH_2PO_4	1.70	0.68 mg/L
2	2.26	2.70	Pd-Sn/Al_2O_3	4.26	CO_2	1.38	<10 μg/L
3	2.26	2.72	Pd/SnO_2	5.03	KH_2PO_4	1.66	<10 μg/L
4	2.26	2.72	Pd/SnO_2	5.36	CO_2	1.65	<10 μg/L
5	11.30	10.0	Pd-Sn/Al_2O_3	5.24	CO_2	2.00	<10 μg/L

Table 3.3 Effect of catalytic reduction on DON using a HPO fraction

Condition	TDN Concentration (μg/L)
Before reduction	420 ± 5
After reduction (3 hours)	
Catalyst Pd-Sn/Al_2O_3	366 ± 37
Catalyst Pd-SnO_2	339 ± 36

Table 3.4 Characteristics of the NOM solutions

n°	Isolate Description	DOC (mg/L)	TDN (mg/L)	UVA_{254} (cm^{-1})	pH
1	Ribou Reservoir TPI	22.50 (±0.52)	2.79	0.620	4.41
2	Seine River TPI	22.18 (±0.20)	2.26	0.867	4.67
3	South Platte River HPO	20.29 (±0.35)	0.55	1.150	4.42
4	Vienne river HPO	22.04 (±0.22)	0.82	2.360	4.02
5	Vienne river TPH	21.68 (±0.12)	1.28	1.190	4.60
6	Gartempe River TPH	18.25 (±0.18)	0.51	0.596	4.25

Table 3.5 Effects of zeolite addition on removal of a Saguaro Lake, Arizona hydrophobic acid NOM isolate

Zeolite Dosage ($V_{Zeolite}/V_{water}$)	DOC Remaining (%)	DON Remaining (%)	UVA_{254} Remaining (%)
0 %	100	100	100
0.5 %	95.2	101.9	99.9
2.0 %	93.5	97.0	95.0
5.0 %	91.6	93.1	96.7

Table 3.6 Summary of dialysis pretreatment method for determining DON concentrations

Step Number	Procedure
#1	Clean 100 dalton dialysis tube; rinse dialysis tubes thoroughly with deionized water over several days
#2	Measure exactly 25 mL of sample (V_0) and place in dialysis tubes; pH 6 to 8 is acceptable
#3	Prepare dialysis system. Dialysis liquid flowrate = 100 mL/min of distilled water. Dialysis reactor volume = 6.7 L; total dialysis tube volume (9 dialysis tubes) is 225 mL; HRT = 67 minutes
#4	Conduct dialysis for 24 hours in a continuous-flow in a dark/covered dialysis system
#5	Remove dialysis bag and transfer using a pipette the entire sample to pre-weighed, clean glass vial. Measure the gravimetric weight of the liquid sample. Calculate the final sample volume (V_{Final}) from the sample weight and a density of 1 cm^3/g. Calculate the swelling ratio: $SR = V_{Final} / V_0$
#6	Analyze the concentration of each nitrogen specie in the dialysis retentate ($N_{i,\ dialysis}$). Conduct TDN analysis by HTO, ammonia analysis by TRAACs, and nitrate & nitrite by ion chromatography. Measure DOC by HTO ($DOC_{i,\ dialysis}$)
#7	Correct concentrations of all nitrogen species (N_i) as needed for ionic strength effects using the swelling ration (SR) calculated for Step #5 and measured nitrogen specie concentration in the dialysis retentate ($N_{i,\ dialysis}$) : $N_i = \{ N_{i,\ dialysis} \} \times \{ SR \}$ $DOC_{final} = \{ DOC_{i,\ dialysis} \} \times \{ SR \}$
#8	Calculate the final DON concentration by subtracting any detectable nitrogen specie concentrations from TDN: $DON_{final} = N_{TDN} - N_{NO3} - N_{NO2} - N_{NH3/NH4}$
#9	Compare DOC with and without dialysis pretreatment. Higher DOC after dialysis pretreatment suggests leaching from improperly cleaned dialysis tube or dirty glassware. Lower DOC after dialysis pretreatment suggests NOM adsorption onto dialysis tube. If DOC concentrations are within 5% of each other with and without dialysis, the accuracy of DOC determination, then no significant loss of NOM occurred during dialysis pretreatment. Verify that the DIN/TDN ratio after dialysis is less than 0.6. If the DOC and DIN/TDN conditions are satisfied, then report the DON_{final} concentration.

Table 3.7 Characteristics of waters used in assessing dialysis pretreatment

Water source	DOC (mgC/L)	DON * (mgN/L)	TDN (mgN/L)	DIN/TDN (mgN/mgN)	SUVA (L/mg-m)	Conductivity (μS/cm)
Arizona Canal water	3.3	0.20	0.34	0.41	1.4	760
Mesa WWTP effluent	5.9	0.65	7.22	0.91	1.6	1079
Trickling WWTP effluent	16.1	1.43	23.7	0.94	2.0	1005
Surface waters (n = 30)	1.7 to 7.0	-0.13 to 0.51	0.24 to 8.8	0.25 to 1.0	1.0 to 4.8	36 to 954
Finished drinking Waters (n = 30)	1.5 to 3.8	-0.42 to 0.54	0.24 to 9.1	0.51 to 1.0	0.5 to 2.2	75 to 938

* Values are calculated using Equation 3.1 without dialysis pretreatment

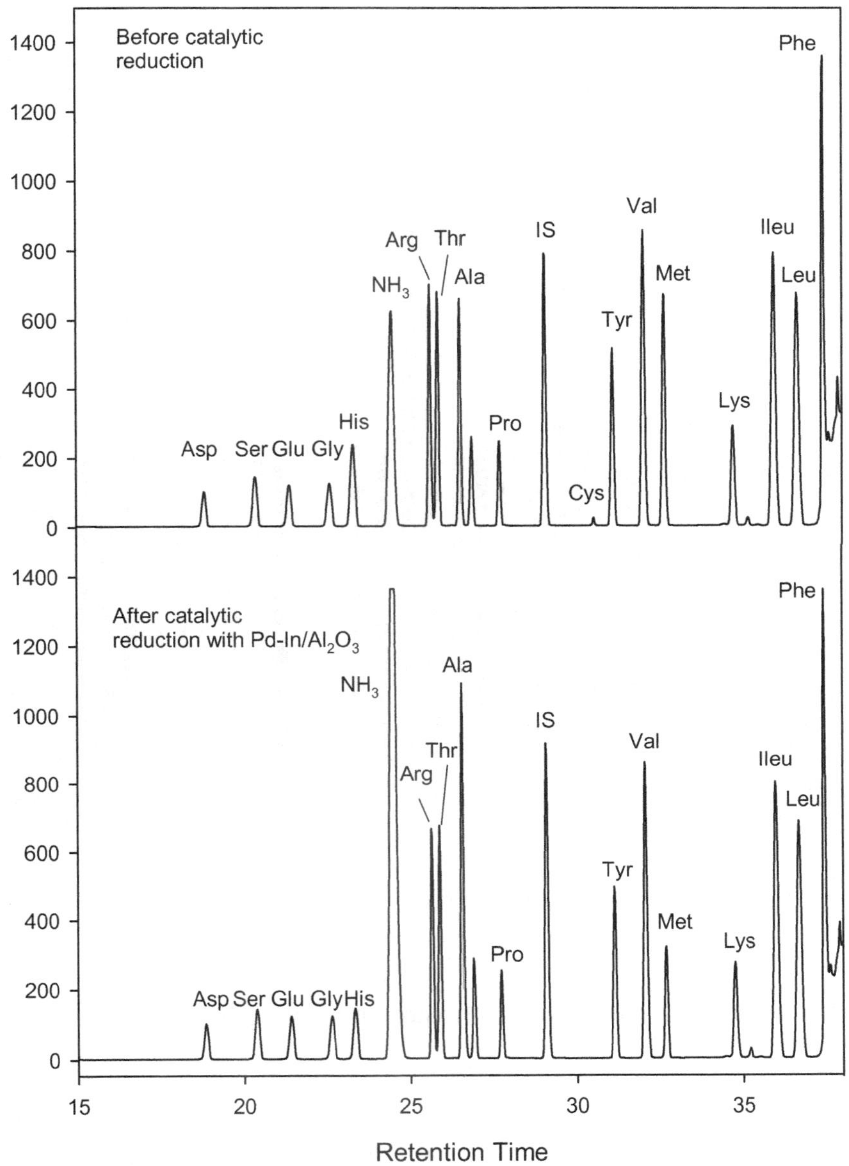

Conditions: catalyst Pd-In/Al_2O_3; 30 minutes reaction time; pH 8.2; room temperature; IS internal standard: α-aminobutyric acid

Figure 3.1 Chromatograms of the amino acids standard solution (10 μM) before and after catalytic reduction

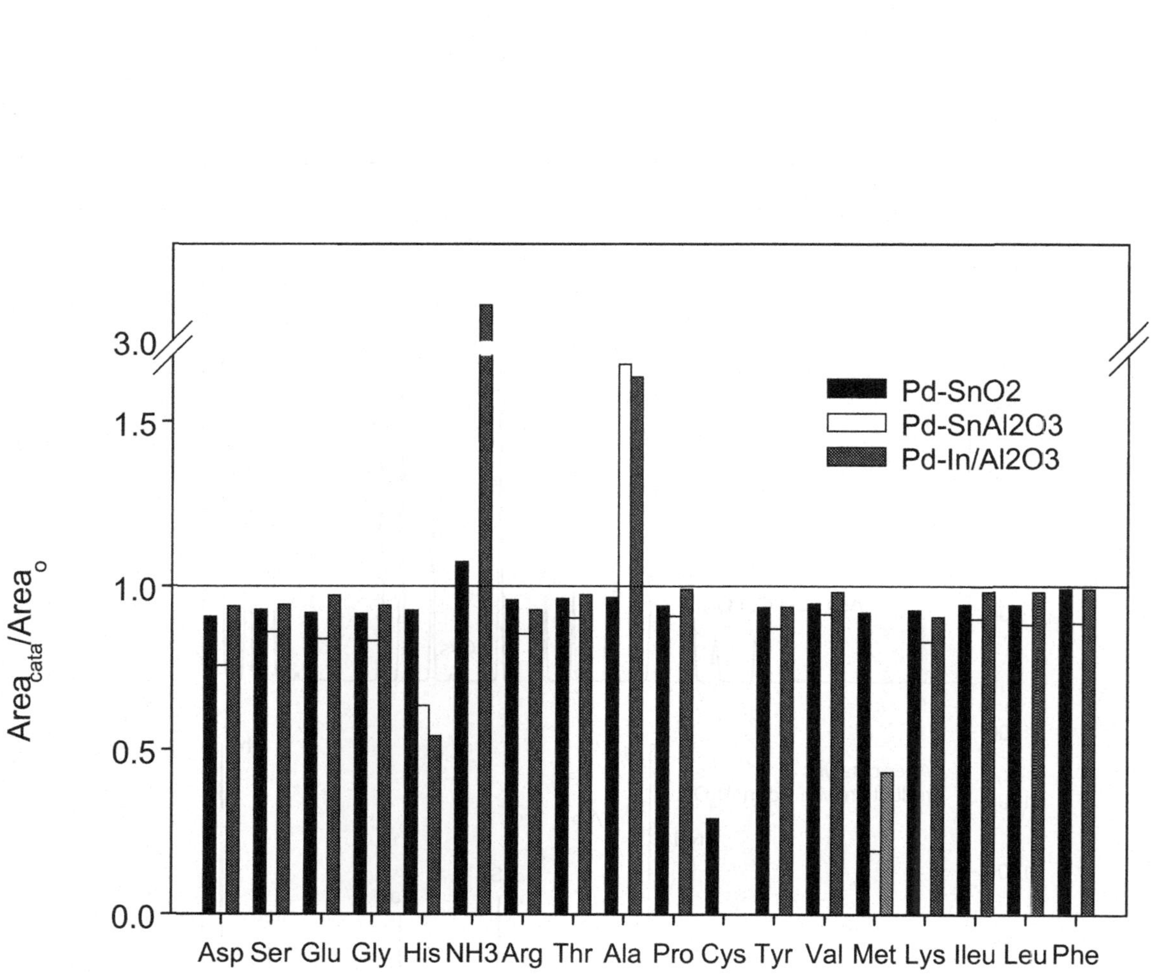

Conditions: 30 minutes catalytic reduction, pH 8.2, room temperature

Figure 3.2 Effect of catalytic reduction on the chromatographic response of a 10 μM amino acid standard

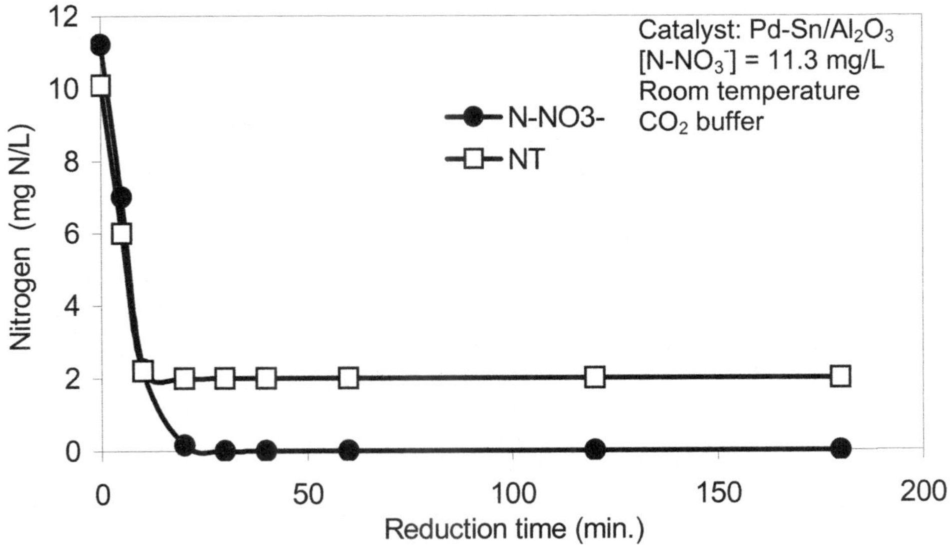

Figure 3.3 Catalytic reduction of a nitrate solution

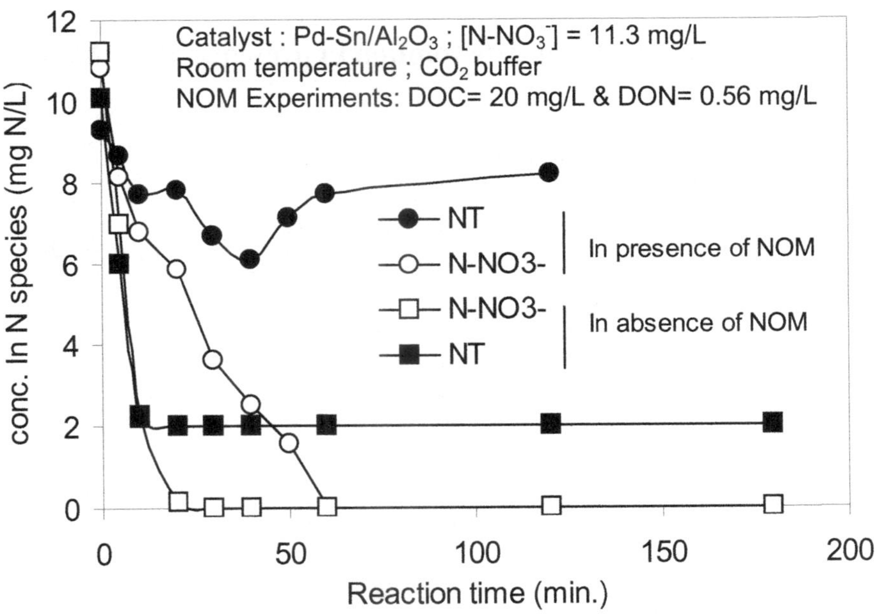

Figure 3.4 Catalytic reduction of a nitrate solution in presence of NOM

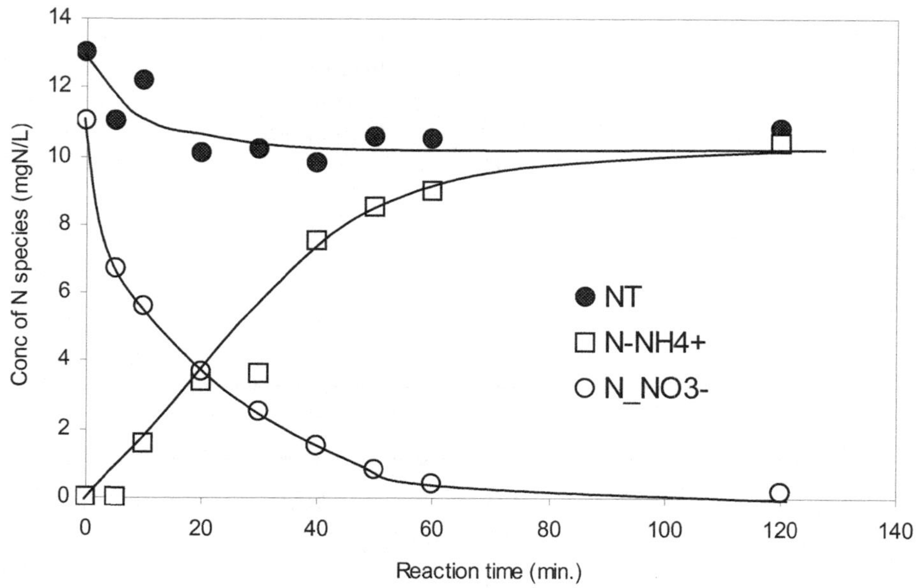

Conditions: Catalyst : Pd-Sn/Al_2O_3 ; [N-NO_3^-] = 11.3 mg/L ; room temperature ; buffer : CO_2; experiments with NOM: DOC = 20 mg/L and DON = 0.56 mg/L

Figure 3.5 Catalytic reduction of a nitrate solution in presence of NOM

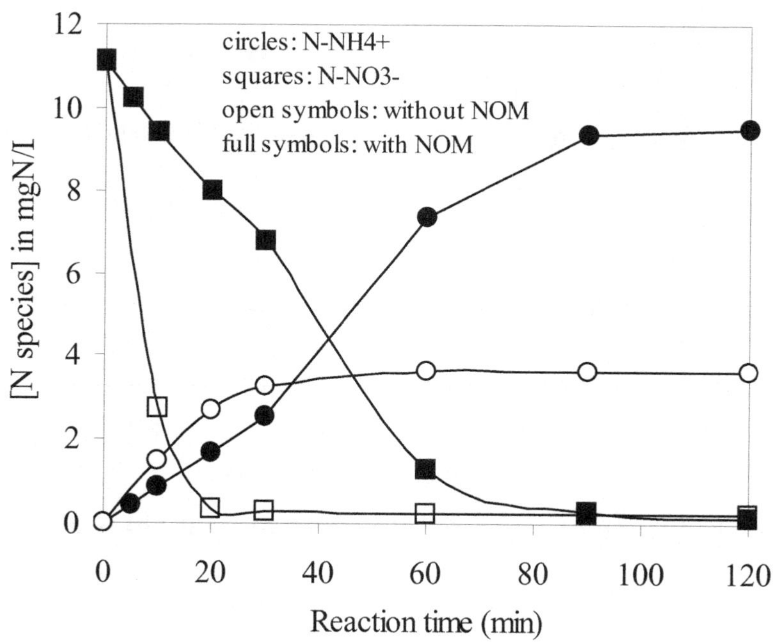

Conditions: Catalyst : Pd-Sn/Al_2O_3 ; [N-NO_3^-] 12 mg/L ; room temperature ; buffer : CO_2; experiments with NOM: DOC = 20 mg/L and DON = 0.67 mg/L

Figure 3.6 Effect of NOM on the catalytic reduction of nitrate and on the formation of ammonia

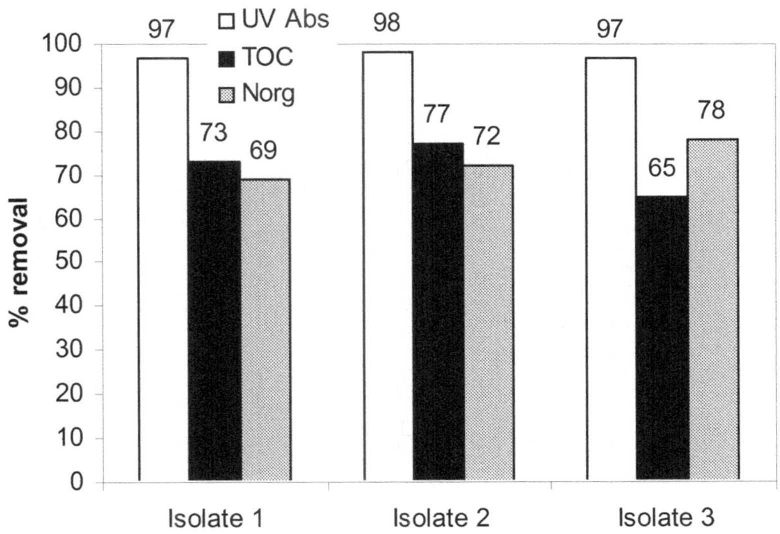

Figure 3.7 Percentage removal of TOC C, org N and UV absorbance for 3 hours contact time with Pd-Sn/Al_2O_3 catalyst

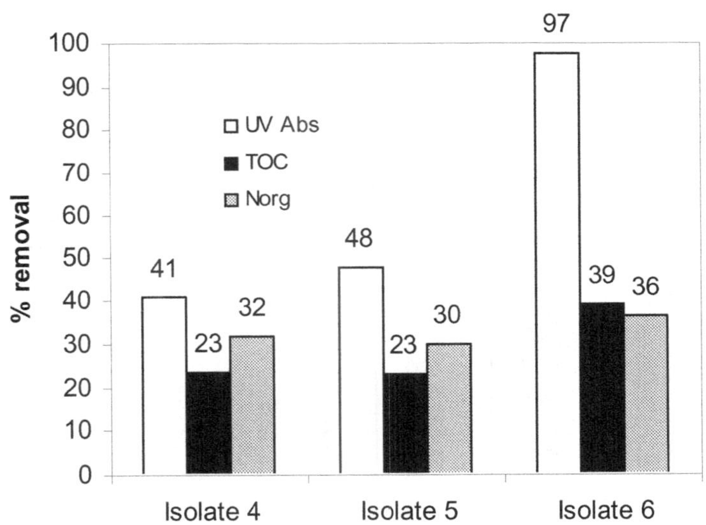

Figure 3.8 Percentage removal of TOC C, org N and UV absorbance for 3 hours contact time with Pd-SnO_2 catalyst

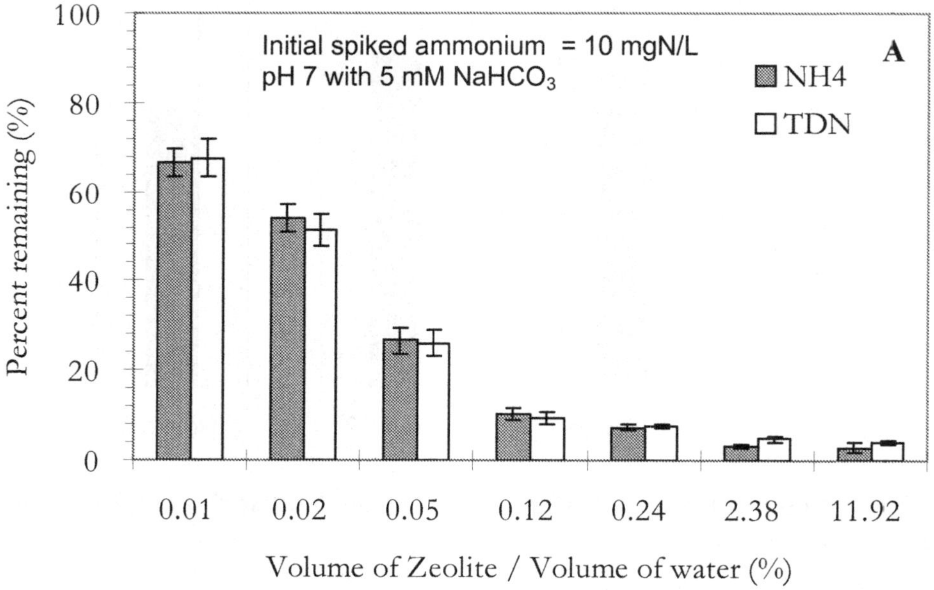

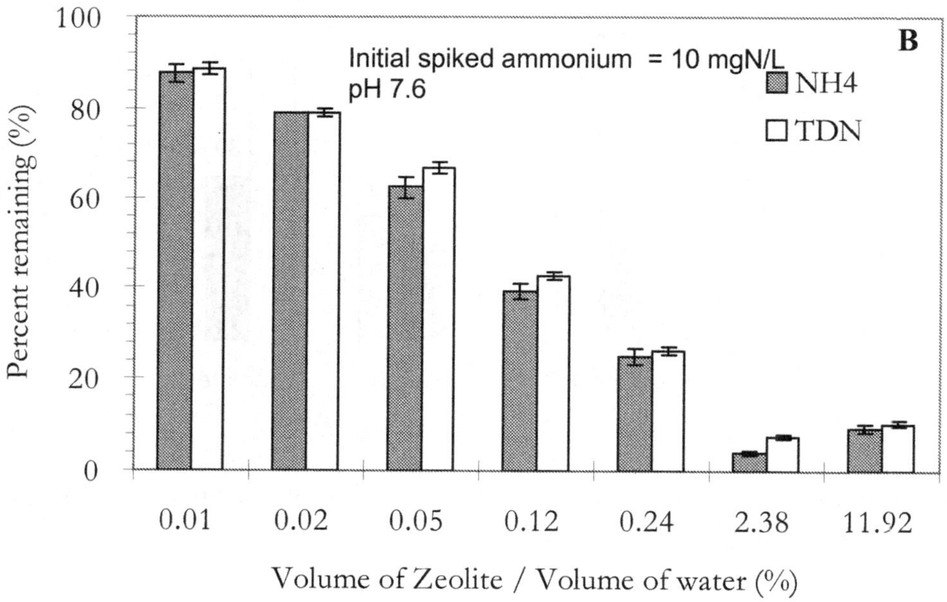

Figure 3.9 Sorption of ammonia on zeolite in (A) Nanopure water and (B) Tempe, Arizona tap water

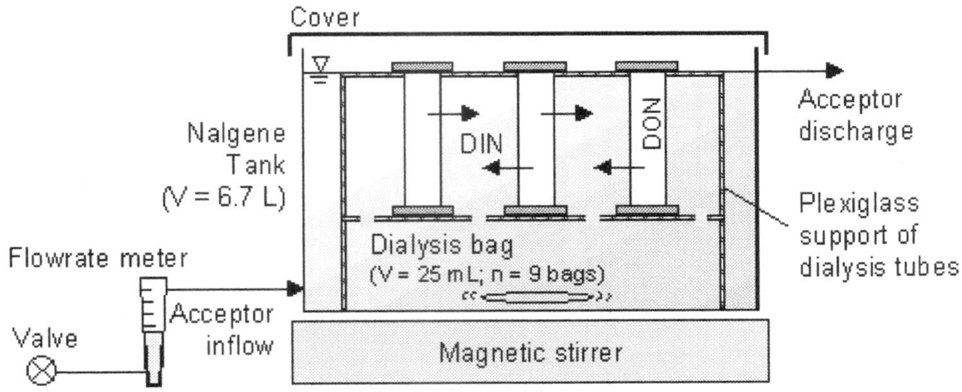

Figure 3.10 Schematic of dialysis pretreatment system for DIN separation from DON

Figure 3.11 Photograph of dialysis pretreatment system

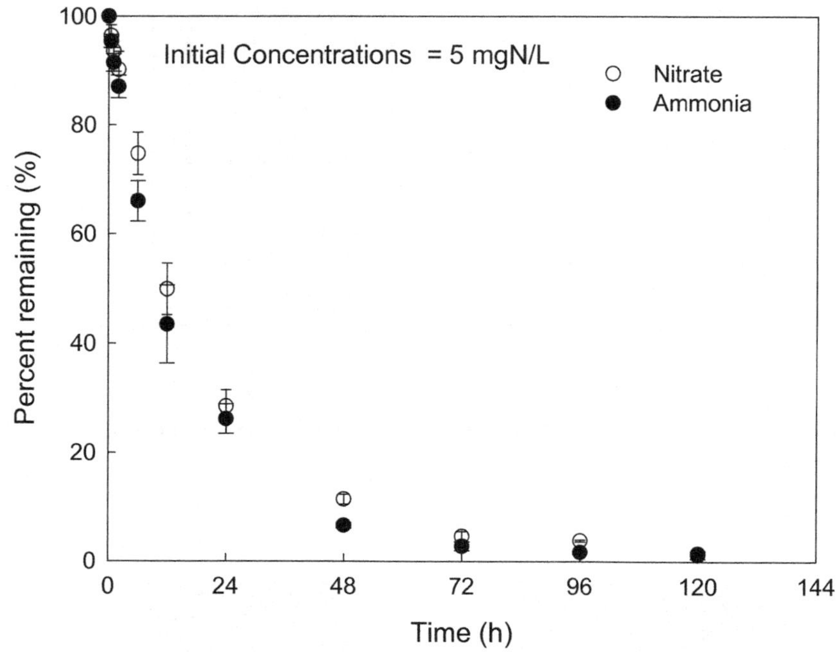

Figure 3.12 Effect of dialysis duration on remaining of nitrate and ammonia

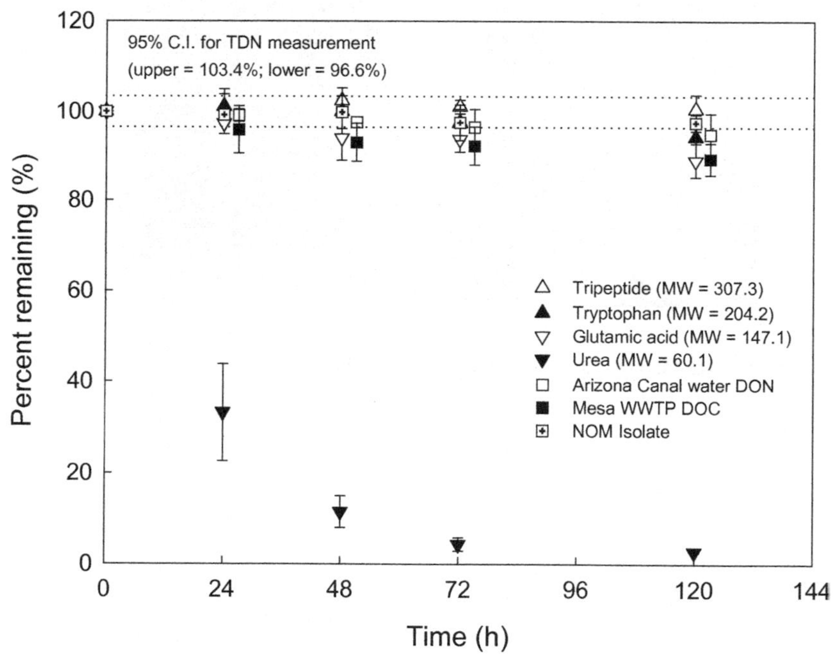

Figure 3.13 Effect of dialysis duration on percentage remaining of organic nitrogen-containing compounds [initial concentration (C_i) = 5 mgN/L], NOM isolate DON [C_i = 1 mgN/L], surface water DON [C_i = 0.2 mgN/L], and treated wastewater DOC [C_i = 6 mgC/L].

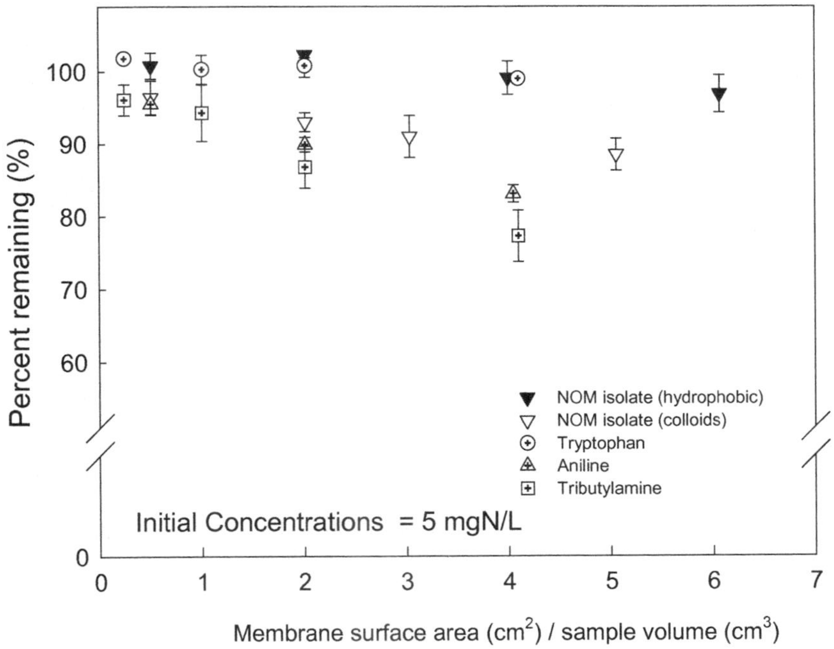

Figure 3.14 Adsorption experiments for organic nitrogen-containing compounds and NOM isolates onto the dialysis membrane.

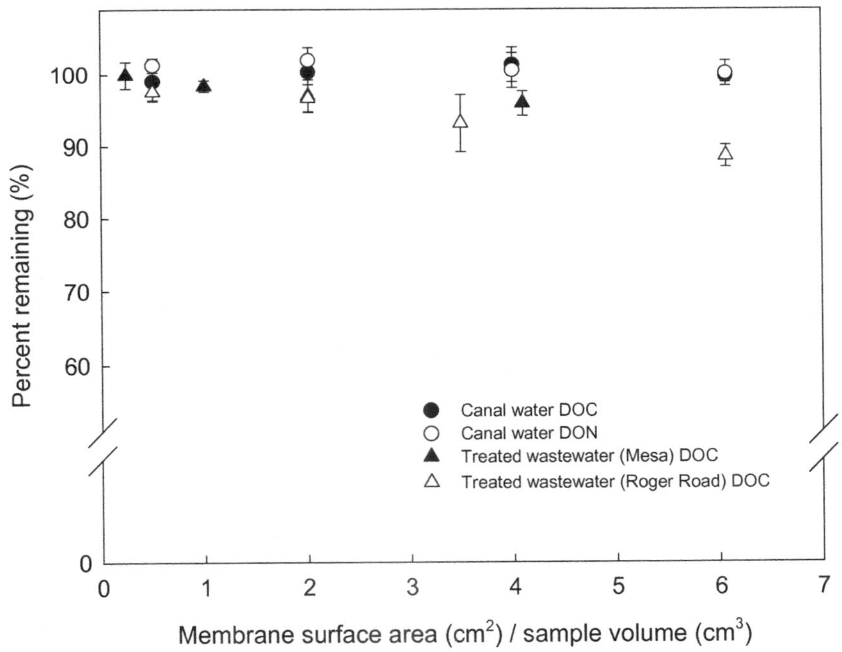

Figure 3.15 Adsorption experiments for bulk waters onto the dialysis membrane.

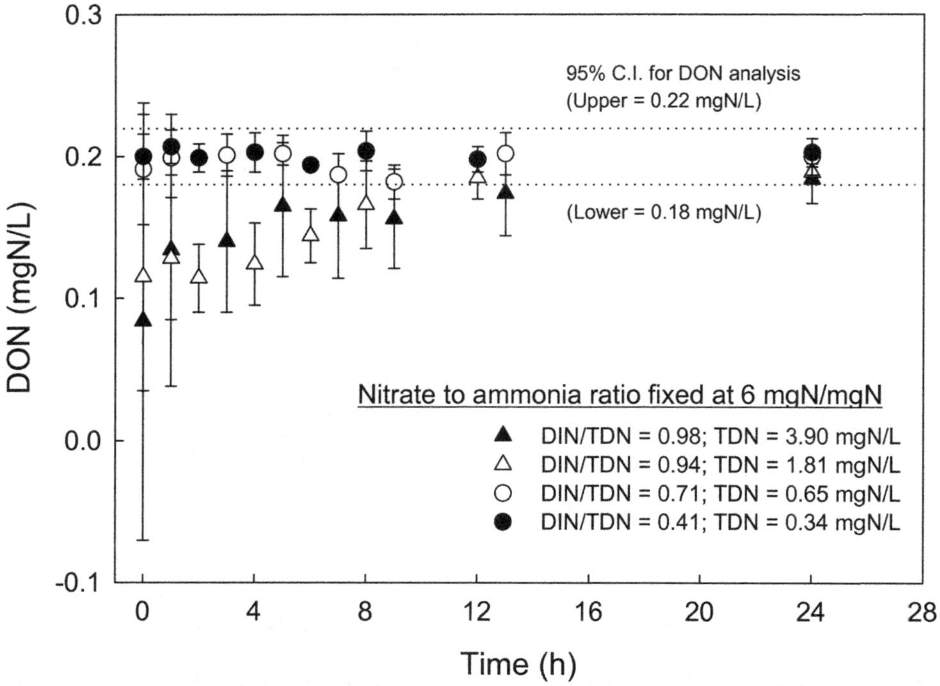

Figure 3.16 Variation of measured DON concentration over time during dialysis of surface water spiked with different DIN/TDN ratio.

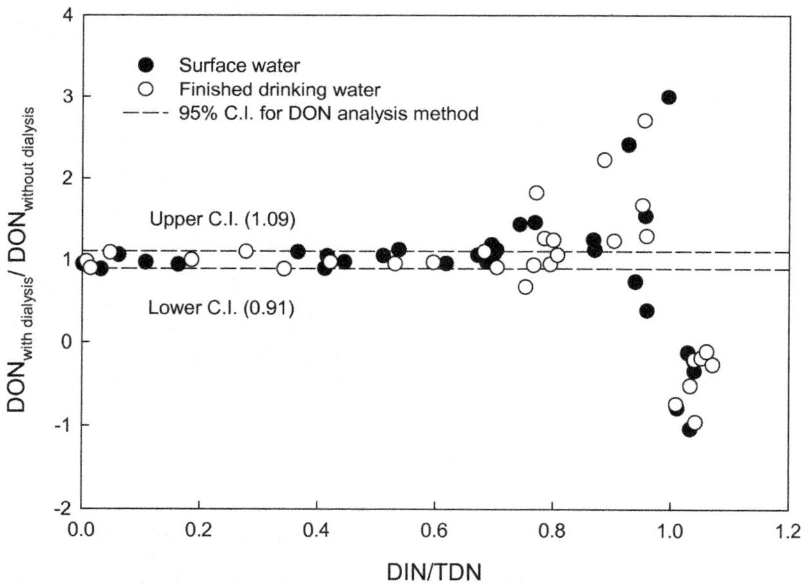

Figure 3.17 Comparison of DON concentrations for surface waters [n = 30] and finished drinking waters [n = 30] with and without the dialysis pretreatment.

CHAPTER 4 DISSOLVED ORGANIC NITROGEN OCCURRENCE

This chapter summarizes DON, nitrate, nitrite, and ammonia occurrence in water supplies, raw/finished drinking waters, and reclaimed wastewaters across the US and France. This represents the first occurrence study of DON in drinking water supplies.

DON SURVEY OF TWENTY EIGHT WATER TREATMENT PLANTS (USA)

Twenty eight water treatment plants (WTPs) in 13 states (Arizona, California, Illinois, Indiana, Kentucky, Massachusetts, Michigan, Missouri, Nevada, Pennsylvania, Texas, Virginia, and Wisconsin) participated in the two sampling campaigns. Figure 2.1 illustrates the geographical distribution of these WTPs. Table 2.1 summarizes individual utility names, locations and reference numbers. Raw water from these facilities include rivers (21 WTPs), reservoirs (5 WTPs), and lakes (2 WTPs). Many of the WTPs were located in the central USA, which has relatively high natural background concentrations total nitrogen occurs(Smith et al. 2003). DIN and DON occurrence from two rounds of utility sampling are described in this section. Information on molecular weight fractionation, UVA absorbance, and fluorescence for a subset of these samples are in a separate chapter dedicated to organic nitrogen characterization and surrogates (Chapter 5). Results from utility samples for organic chloramine formation upon addition of disinfectants are also presented later (Chapter 7).

Inorganic Nitrogen Species Occurrence

The nitrogen specie distributions are summarized in Table 4.1 for WTP raw waters collected during two seasonal utility sampling campaigns. The average TDN concentration of WTP raw waters was 1.809 mgN/L (median = 0.923 mgN/L), and 70% of the waters had TDN less than 2 mgN/L (Figure 4.1). Nitrate was the dominant inorganic nitrogen species, on average accounting for 90% of the DIN. Nitrate concentrations for each WTP raw water are shown in Figure 4.2. Nitrite concentrations in most samples were below the MDL of 0.005 mgN/L. In most cases ammonia plus ammonium ion concentrations were less than 0.1 mgN/L (Figure 4.3). The average DIN/TDN ratio for the WTP raw waters was 0.76 mgN/L, which is greater than the maximum DIN/TDN ratio of 0.6 mgN/mgN recommended for analysis without dialysis pretreatment (see Chapter 3).

Organic Nitrogen Occurrence in Raw and Finished Waters

Since the average DIN/TDN ratio was >0.6 mgN/mgN (see previous section), dialysis pretreatment was used prior to DON determination as follows: 1) DON without dialysis pretreatment is reported for samples with DIN/TDN ratios ≤ 0.6 mgN/mgN, or 2) DON with dialysis pretreatment is reported for samples with DIN/TDN > 0.6 mgN/mgN. DOC data is reported for samples without dialysis pretreatment, unless specifically stated otherwise. Table 4.2 summarizes the DOC and DON concentration data from the utility survey. The ratio of DOC/DON in the raw waters from the utility surveys averaged 18 mgC/mgN, ranging from 7 to 32 mgC/mgN, for samples with >0.06 mgN/L. For six river samples with low DON concentrations (< 0.06 mgN/L), the DOC/DON ratios were much higher (60 to 135 mgC/mgN). The average DON concentration in raw water was 0.19 mgN/L. This value is at or just below the detection limit of TKN methods. Therefore, TKN methods are probably not suitable for quantifying changes in DON across most water treatment plants.

DON concentrations in raw and finished waters from each utility, and the percentage DON removed, are plotted in Figure 4.4. The data was plotted from lowest to highest DON in raw water observed in the May-June 2003 survey at each WTP. DON concentrations collected during the second utility sampling campaign did not show any consistent pattern, and were higher, equal, or lower than the data collected earlier in the year. Many factors could be responsible for differences in DON between the two surveys (e.g., hydrologic variations, upstream inputs, algae activity, etc.). However, most of the utilities with more than a 50% difference in DON between the two seasons had greater than an 8 ^{o}C water temperature difference between the two sampling campaign dates, and thus it was hypothesized that temperature might have a strong influence on DON concentrations. Overall, DON showed only a weak statistical correlation with temperature (higher DON with higher temperature). However, for samples collected within a similar geographic area (seven utilities in Illinois) a positive linear correlation between DON and water temperature was observed (R^2=0.85). Separate monitoring for three water supply reservoirs in central Arizona also exhibited a positive trending relationship between DON concentration and water temperature. Temperature may be a bulk integrator for algae activity and hydrologic variability.

The percentage DON removed across the WTPs sampled averaged 20%, while DOC and UVA_{254} removals of 29% and 55% were observed, respectively. SUVA values of the raw waters ranged from 1.0 to 4.3 $(mgC/L)^{-1}m^{-1}$. Higher raw water SUVA values correlated with higher percentages of DOC removal (r=0.61) but not with DON removal (r =0.13). DOC and DON percentage removals on paired samples at different WTPs did not exhibit statistically significant correlations (Figure 4.5). Only 10% of the WTP samples had higher DON in finished waters. DON and SUVA did not exhibit a statistical dependence (Figure 4.6). It is generally accepted that coagulation preferentially removes material with higher SUVA values, and is consistent with the high 55% UVA_{254} removal. Since DON and SUVA do not correlate, this suggests that nitrogenous organic material is associated with lower SUVA fractions of NOM that are more poorly removed during coagulation.

Most of the WTPs used polymers (21 of 28 utilities), and 18 of those 21 using polymers employed cationic polymers with dosages ranging from 0.5 to 3.5 mg/L. On average utilities using polymers averaged 23% DON removal, while those without polymers had less removal (9% DON removal). The same pattern was observed during both sampling campaigns. It is unclear if the use of cationic polymers were the direct cause of the higher DON removal, or just coincidental. Cationic polymers contain organic nitrogen, yet appear to improve DON removal. Based upon this observation bench-scale experiments were conducted to further investigate a hypothesis that cationic polymer addition improves DON removal during coagulation (see Chapter 6).

Nearly 40% of the WTPs sampled reported that their filters were operated biologically. Since bacteria in filters could convert inorganic nitrogen into DON, it was hypothesized that biofiltration may increase DON concentrations across the treatment train. However, only a small difference was observed between DON removal at WTPs with biofiltration (22% DON removal) and those without (19% DON removal). Four of the eleven utilities with biofiltration also used ozone. DON removal for utilities with ozone (34% DON removal) was slightly higher than those without ozone (16% DON removal). A summary of all sample data from the utility survey is provided in Table 4.3.

SPATIAL AND TEMPORAL SAMPLING IN SURFACE WATERS AND RECLAIMED WASTEWATERS (USA)

Several drinking water sources were monitored, or data provided, to evaluate temporal changes in DON concentrations, or spatial variability within a source. Figure 4.7 is an example of this type of data for DON concentrations collected at a depth of ~10m below the water surface for three reservoirs in central Arizona. For all three reservoirs DON concentrations increase during the summer months with the highest concentrations between August and September, before decreasing in concentration during the winter months. During the winter the DIN/TDN ratio is approximately 0.5, but during the summer nearly 100% of the TDN is present as DON. The warm summer months correlate well with high algae productivity, and it was hypothesized that the algae produce soluble microbial products enriched in organic nitrogen compounds (e.g., chlorophyll, proteins, colloidal cell wall material, etc.).

Figures 4.8 through 4.10 illustrate DON and DOC/DON ratios for three different locations over multiple week periods. Only for Figure 4.10 was significant variation observed, and was associated with a shift in hydrologic conditions. Along with other sampling efforts, larger watersheds tend to exhibit relatively constant DON levels over the period of weeks. However, some source waters such as rivers can experience much greater spatial variability. Figure 4.11 is one example of this for the South Platte River through the Denver metro area. DON concentrations increased from an upstream site (site #1) through the metropolitan region, and was affected by treated wastewater discharges into the river. Reservoirs can also be impacted by wastewater discharges. An example of this is Lake Mead on the Colorado River which can be impacted by wastewater conveyed through the Las Vegas Wash into Lake Mead. A map of locations sampled by SNWA was presented in Figure 2.3, and representative data presented in Figure 4.12. DON was calculated as TKN minus ammonia; some negative values were observed in the database. As water moves further away from the source of the wastewater, DON concentrations decreased. In Las Vegas Bay near the mouth of the Las Vegas Wash near surface DON concentrations were highest and decreased with depth. This could be due to thermal stratification of inflowing wastewater or algae activity. In the deeper sections of the reservoir, DON concentrations were nearly uniform with depth.

Temporal sampling was also conducted at a number of water treatment plants in raw and finished waters. Examples from Arizona included an ozone/GAC WTP (Greenway WTP) and a direct filtration WTP (Union Hills WTP) (Figures 4.13 and 4.14). In both cases seasonal variations in raw water DON exist, and are influenced by algae activity in upstream reservoirs and other hydrologic/engineered operations (e.g., groundwater is pumped into the canal supply Greenway WTP with a combination of ground and surface water). In greater than 95% of the samples, finished water DON was less than raw water DON. However, it was observed during sampling at other facilities that DON could increase across specific unit processes, while decreasing across an entire WTP. One example of this phenomena is from a WTP in Illinois. DON decreased from 0.36 mgN/L in raw water during preoxidation and coagulation (0.23 mgN/L), but increased across the biological filters (0.38 mgN/L). Final disinfection appeared to oxidize the DON produced in the filters (final DON was 0.18 mgN/L), or may have been lower because sampling was not flow-paced across the WTP. For utilities using chloramine, some residual free ammonia is usually present. In these cases we often saw a slight decrease in free ammonia and increase in DON across filtration.

Samples were collected from three wastewater treatment plants and at associated reclaimed wastewater facilities. These samples were collected and analyzed prior to completion of the dialysis pretreatment method, and therefore samples were not dialysized. At the Phoenix, Arizona 91st Avenue WWTP (partial denitrification; DOC ~ 3.75 mg/L; DIN/TDN = 0.8 mgN/mgN) a portion of the wastewater is discharged to a wetland treatment system. The DON entering the wetland was 2.8 mgN/L and 1.4 mgN/L exiting. Samples collected from shallow wells below the "leaky" wetland contained DON (0.3 to 0.6 mgN/L), and the DON concentrations decreased with depth. At a trickling filter WWTP in Tucson, Arizona the DIN/TDN ratio was >1.0 (TDN = 19 mgN/L; DOC ~ 15 mg/L) and thus DON could not be reported. This exemplifies the need for dialysis pretreatment strategy. However, the DON in a shallow monitoring well below this recharged wastewater (surface spreading basins) the DON was 2.1 mgN/L. This DON concentration is substantially higher than the native groundwater (<0.1 mgN/L). The Mesa, Arizona WWTP includes nitrification and denitrification processes (DOC ~ 4 mg/L; DIN/TDN = 0.9 mgN/mgN) and has an effluent DON of 1.1 mgN/L. After application of the effluent to spreading basins (i.e., soil aquifer treatment) the DON in a series of monitoring wells decreases with increasing travel time away from the spreading basins and was below 0.1 mgN/L after the equivalent of 6 months of travel time.

It is important to note that the above WWTP DON values are based upon measurement without dialysis and at the high DIN/TDN ratios the accuracy of DON values are low for DIN/TDN ratios above 0.6 mgN/mgN (see Chapter 3). Therefore, additional measurements at the Phoenix 91st Avenue WWTP facility were made later in the project using dialysis pretreatment. Primary effluent had a DON concentration of 5.25 mgN/L (DOC = 25 mg/L). Secondary effluent had a DON concentration of 1.34 mgN/L (DOC = 4.76 mg/L). In addition, a molecular weight fractionation of the samples were conducted. For the primary effluent 67% of the DON had a molecular weight of < 1000 dalton, with 27% >10,000 dalton. The secondary effluent 80% of the DON had a molecular weight of < 1000 dalton, but only 7% >10,000 dalton. While secondary treatment removed all molecular weight fractions of DON (and DOC), it appeared to effectively reduce the higher molecular weight fraction preferentially.

SAMPLING IN FRANCE

Watershed Sampling

Three river systems in France (Sèvre Niortaise River, the Marne River and the Vienne River) were studied (Figure 4.15). Table 4.4 summarizes the average DOC, TDN, and DIN for the three rivers sampled between March and June 2004. All three rivers had very high DIN/TDN ratios, thus requiring dialysis pretreatment in order to accurately quantify DON concentrations.

Samples were collected downstream and upstream of Saint Maixent and Annet sur Marne to evaluate the impact of domestic and industrial wastewater effluents. Ammonium and nitrite were always found below the MDLs. Table 4.5 gives the DOC, DIN and TN values determined before dialysis. DON values are also given and were calculated by subtracting DIN to TDN after dialysis. Results show that Sèvre Niortaise River waters are characterized by low DOC (< 1mgC/L) and high DIN values (>9.8mgN/L). For Sevre Niortaise River water, DIN values were found higher than the TDN values, which made impossible the determination of DON without dialysis. Vienne and Marne river waters exhibit higher DOC content from 2.0 mgC/L to 3.5 mgC/L but lower DIN values (< 4.6 mgN/L). After dialysis, DON values range between 0.18 µgN/L to 0.57 µgN/L. DOC/DON ratios are unexpectedly very low for Sevre Niortaise River with values ranging from 2.4 to 3.6 mgC/mgN. Results show also that the treated waste water from the abattoir and the different cities along the river have no impact on the DOC or DON values. Only a small increase of DON values between April and June for the Marne River, which may be attributed to the growth of algae.

Surface waters were also sampled in the Cebron reservoir basin (Figure 4.16) during 2 surveys in mid-May and mid-June 2004. The figure presents the hydrographic map of the Cébron reservoir basin with the locations of the 18 sample points. These sample points are located on the reservoir, the four creeks connected to the reservoir and different pools and ditches. This rural area is mainly used for extensive farming (cattle, sheep). The sources of domestic sewage are limited to the village of Viennay (one waste water treatment plant) and Adilly (direct sewage run-off from inhabitations). These surface waters is characterized by high DOC concentrations from 7 to 30 mgC/L. Nitrate concentrations were found from 0 to 3 mgN/L. Table 4.6 summarizes data from sampling in May. DON was measured without dialysis pretreatment. Sampling was repeated in June 2004, but some ditches were dry and samples were not collected. Figure 4.17 and 4.18 illustrates nitrate DON data for the sites in the two months of sampling. DON values in May 2004 survey range between 0.48 mgN/L to 1.75 mgN/L. The DOC/DON ratios range between 8.0 for a creek impacted by a WWTP (sample n° 10) and 21.3 for water from the Cebron reservoir (sample n°17). The evolution of nitrate and DON from May 2004 to June 2004 depicted in Figures 4.17 and 4.18 show a drastic reduction of nitrate from May to June except for sample n° 10, whose nitrate (and also DON) content is highly dependent on the effluent of the Viennay WWTP. Simultaneously with changing nitrate levels, DON values significantly increase from May to June. Both results can be explained by the growth of algae during summer (uptake of nitrate and formation of autochthonous NOM).

For the Cebron reservoir basin a comparison between DOC and DON concentrations is depicted in Figure 4.19. For most of samples a good correlation can be found, with a mean DOC/DON ratio of 13.0. Samples 10 and 13 exhibit lower DOC/DON ratios and the DOC/DON ratio depend on the sampling period. These lower values and the higher variability can be explained by the influence of WWTP effluent for sample n°10 and by sewage run-off coming from several inhabitations in Adilly village (n°13).

Water Treatment Plant Sampling

Two drinking water utilities were studied. The Saint Maixent WTP receives water from a water supply reservoir and employs coagulation/flocculation (~ 20 mg/L ferric chloride) followed by flotation, sand filtration, GAC adsorption, and chlorine disinfection. Figure 4.20 illustrates the change in DON and DOC concentrations across this WTP series of processes. Overall the DOC and DON removals were 64% and 53%, respectively. The coagulation/flotation process accounted for most of the DON removal, while each of the sequential processes achieved incremental DOC removal. It is possible that DON from the source water was removed in the sand and GAC systems, but that biological activity also produce "new" soluble microbial products containing DON.

Sampling in two different months at a WTP on the Marne River (France) was also conducted (Table 4.7, Figures 4.21 and 4.22). The treatment train consists of coagulation/ flocculation, solids separation, sand/anthracite filtration, ozonation, and chlorination. Minimal DOC removal was observed, especially during the June sample. DON concentrations increased during coagulation, possibly due to usage of cationic polymers that contain organic nitrogen, and further bench-scale work was conducted to evaluate this (see Chapter 6). Filtration may have removed carry-over floc containing residual cationic polymer. Overall approximately 25% of the DON was removed across this WTP.

SUMMARY

Dialysis pretreatment with a high temperature catalytic oxidation technique for nitrogen analysis (Shimadzu TOC-V) was used for water treatment plant sampling since the average DIN/TDN was greater than 0.6 mgN/mgN. The following observations were made from a study of two sampling campaigns conducted at 28 WTPs in the US:

- The mean DON concentration for 28 water treatment plants in the US was 0.19 mgN/L for raw waters and 0.15 mgN/L for finished waters. The DON concentrations ranged from <0.05 to 0.45 mgN/L.
- The mean concentrations were below the 0.2 mgN/L MDL for Total Kjeldahl nitrogen measurements. This suggests that other methods, such as the high temperature catalytic oxidation technique used herein, with lower MDLs are required for wide-spread use in the drinking water industry.
- The average DOC/DON ratio for the raw waters was 18mgC/mgN.
- Average DON removal across the 28 WTPs was 20%, which was lower than the average DOC removal (29%) or UVA_{254} removal (55%). WTPs using polymers had approximately a 9% higher DON removal than WTPs not reporting use of polymer during coagulation.
- Seasonally DON concentrations increased in the epilimnion of water supply reservoirs, during warmer water temperatures and periods of algae activity that produced soluble microbial products. In some reservoirs the DIN/TDN ratio effectively decreased to zero, as all the dissolved nitrogen became incorporated as DON.
- Within common watersheds or similar hydrologic settings, higher DON concentrations correlated with increasing water temperatures. This suggests temperature may be an indicator for DON variability. This observation could result from algae and bacteria activity, or changes in hydrologic flowpaths.

For two WTPs in the US and France, most of the DON occurred during coagulation. For WTPs with granular activated carbon, additional removal was observed. For a WTP in France with coagulation, filtration, and ozonation the DON removal was 25% and comparable with DON removals observed at the US WTPs. For a second WTP in France with additional treatment processes (flotation, GAC), higher DON removals (53%) were observed. At a few US WTP that practiced chloramination prior to filtration (e.g., excess free ammonia present), a transformation of free ammonia to DON during filtration was observed and attributed to bacterial processes.

In both the US and France discharging wastewater into surface waters (reservoirs or rivers) increases DON concentrations. Wastewater contains higher DON than unimpacted surface waters. Furthermore the addition of nutrients in wastewater effluents increases the potential for algae production, and subsequent formation of soluble microbial products that contain DON. Discharge of wastewater into the subsurface (i.e., soil aquifer treatment) decreases the DON and DOC concentrations. The magnitude of these reductions depends upon travel time.

Table 4.1 Summary of nitrogen specie concentration without dialysis for WTP raw waters collected over two utility sampling campaigns [n=56]

Percentile Ranking	TDN (mgN/L)	Nitrate (mgN/L)	Ammonia plus Ammonium Ion (mgN/L)
10th	0.29	0.08	<0.005
25th	0.45	0.29	<0.005
Mean	0.923	1.63	0.032
Median	1.809	0.80	0.010
75th	2.27	2.23	0.040
90th	5.33	5.45	0.080

Table 4.2 Summary of DOC and DON in raw and finished water from the two utility sampling campaigns [n=56].

Percentile Ranking	DOC (mg/L)		DON (mgN/L)	
	Raw	Finished	Raw	Finished
10th	1.90	1.63	0.07	0.04
25th	2.56	1.92	0.12	0.10
Mean	3.44	2.44	0.19	0.15
Median	3.05	2.44	0.19	0.14
75th	4.19	2.84	0.23	0.18
90th	5.32	3.52	0.31	0.25

Table 4.3 Results from utility sampling

Utility ID #	Sampling Date	Raw water (mgN/L) TDN	NO_3^-	NH_4^+	DON w/dialysis pretreatment	Finished water (mgN/L) TDN	NO_3^-	NH_4^+	DON w/dialysis pretreatment	Percent change in DON
1	May 2003	0.887	0.793	0.010	0.119	0.965	0.913	0.450	0.039	67%
	Nov 2003	0.907	0.801	0.008	0.098	0.965	0.913	0.000	0.054	45%
2	May 2003	1.468	1.399	0.010	0.159	0.664	0.583	0.250	0.100	37%
	Nov 2003	1.221	1.011	0.045	0.165	1.519	1.365	0.034	0.121	27%
3	May 2003	0.341	0.185	0.000	0.160	0.350	0.224	0.030	0.147	8%
	Oct 2003	0.286	0.005	0.040	0.241	0.215	0.017	0.033	0.155	36%
4	May 2003	0.366	0.180	0.020	0.176	0.371	0.197	0.000	0.164	7%
	Nov 2003	0.278	0.037	0.073	0.168	0.258	0.018	0.079	0.165	2%
5	May 2003	0.239	0.110	0.000	0.151	0.241	0.122	0.000	0.149	1%
	Nov 2003	0.298	0.058	0.045	0.196	0.247	0.057	0.039	0.160	18%
6	May 2003	0.392	0.286	0.000	0.121	0.323	0.239	0.000	0.118	3%
	Nov 2003	0.402	0.296	0.000	0.106	0.313	0.239	0.000	0.079	25%
7	Jun 2003	0.329	0.084	0.037	0.191	0.792	0.113	0.422	0.205	-7%
	Oct 2003	0.249	0.024	0.003	0.216	0.709	0.183	0.316	0.192	11%
8	May 2003	0.509	0.323	0.003	0.160	0.515	0.363	0.000	0.141	12%
	Oct 2003	0.497	0.345	0.024	0.183	0.541	0.417	0.000	0.227	-24%
9	May 2003	4.861	4.370	0.003	0.246	5.576	5.034	0.003	0.226	8%
	Dec 2003	2.752	2.596	0.044	0.044	3.618	2.551	1.098	0.023	48%
10	May 2003	3.008	3.077	0.060	0.296	4.079	3.461	0.680	0.146	51%
	Oct 2003	1.072	0.740	0.005	0.387	1.897	0.797	0.720	0.471	-22%
11	Apr 2003	1.908	1.614	0.000	0.294	2.666	1.785	0.800	0.362	-23%
	Dec 2003	2.784	2.715	0.055	0.042	3.646	2.783	1.006	0.030	29%
12	Apr 2003	2.485	2.351	0.030	0.307	2.476	2.246	0.010	0.186	39%
	Dec 2003	6.275	6.263	0.080	0.053	5.118	5.422	0.010	0.034	37%
13	Jun 2003	8.749	8.522	0.080	0.388	9.107	8.532	0.390	0.196	49%
	Oct 2003	0.932	0.628	0.027	0.310	1.030	0.451	0.532	0.127	59%
14	Jun 2003	8.078	7.559	0.009	0.432	5.259	4.744	0.000	0.348	19%
	Dec 2003	6.540	6.786	0.013	0.087	5.125	5.329	0.002	0.082	6%
15	Apr 2003	5.948	5.827	0.000	0.055	5.044	5.449	0.020	0.117	-113%
	Oct 2003	5.332	5.451	0.038	0.019	3.061	3.196	0.024	0.029	-53%

(continued)

Table 4.3 (continued)

Utility ID #	Sampling Date	Raw water (mgN/L) TDN	NO_3^-	NH_4^+	DON w/dialysis pretreatment	Finished water (mgN/L) TDN	NO_3^-	NH_4^+	DON w/dialysis pretreatment	Percent change in DON
16	Apr 2003	3.042	3.059	0.000	0.187	2.841	2.890	0.015	0.095	50%
	Dec 2003	2.013	1.919	0.005	0.137	1.749	1.663	0.000	0.144	-5%
17	May 2003	0.893	0.674	0.000	0.205	0.625	0.538	0.110	0.095	54%
	Dec 2003	2.270	2.258	0.018	0.023	2.173	1.937	0.391	0.040	-75%
18	Apr 2003	0.914	0.910	0.000	0.105	1.357	1.045	0.450	0.097	7%
	Oct 2003	1.399	1.190	0.027	0.204	1.768	1.142	0.552	0.095	53%
19	May 2003	0.254	0.175	0.000	0.087	0.547	0.175	0.200	0.143	-64%
	Oct 2003	0.188	0.063	0.038	0.104	0.421	0.057	0.316	0.106	-3%
20	Apr 2003	0.915	0.595	0.025	0.368	1.338	0.692	0.500	0.178	52%
	Oct 2003	0.748	0.331	0.002	0.405	1.132	0.392	0.476	0.248	39%
21	Apr 2003	2.348	2.228	0.000	0.131	2.448	2.217	0.100	0.153	-17%
	Oct 2003	2.064	2.123	0.010	0.071	2.471	2.161	0.392	0.042	41%
22	Apr 2003	1.190	1.082	0.040	0.225	1.904	1.161	0.645	0.113	50%
	Dec 2003	1.145	0.968	0.093	0.202	1.491	0.966	0.534	0.113	44%
23	Apr 2003	1.594	0.971	0.455	0.233	1.648	0.730	0.525	0.351	-51%
	Dec 2003	1.320	1.044	0.102	0.216	1.792	0.995	0.624	0.213	2%
24	May 2003	0.621	0.515	0.010	0.187	0.595	0.533	0.081	0.111	40%
	Oct 2003	0.619	0.476	0.000	0.208	0.600	0.471	0.000	0.163	22%
25	Jun 2003	1.871	1.722	0.012	0.191	3.092	2.660	0.422	0.142	26%
	Jan 2004	3.829	3.434	0.162	0.173	3.988	3.399	0.755	0.158	9%
26	May 2003	0.449	0.235	0.000	0.221	1.117	0.247	0.675	0.156	30%
	Oct 2003	0.593	0.195	0.049	0.288	1.188	0.253	0.692	0.231	20%
27	Apr 2003	0.292	0.075	0.000	0.206	0.933	0.075	0.660	0.114	45%
	Oct 2003	0.483	0.177	0.000	0.335	1.048	0.073	0.716	0.175	48%
28	May 2003	0.403	0.291	0.010	0.115	0.532	0.283	0.160	0.101	12%
	Nov 2003	0.469	0.328	0.000	0.151	0.685	0.353	0.200	0.140	7%
min		**0.188**	**0.005**	**0.000**	**0.019**	**0.215**	**0.017**	**0.000**	**0.023**	**-113%**
Mean		**1.809**	**1.633**	**0.032**	**0.188**	**1.896**	**1.515**	**0.306**	**0.148**	**15%**
Max		**8.749**	**8.522**	**0.455**	**0.432**	**9.107**	**8.532**	**1.098**	**0.471**	**67%**

Table 4.4 Average DOC, TDN, and DIN values for the three rivers studied

	DOC (mg/L)	TDN (mgN/L)	DIN (mgN/L)
Marne River	2.1	4.4	4.0
Vienne River	3.3	2.0	1.9
Sèvre Niortaise River	0.9	10	10

Table 4.5 DOC, TDN, DIN values [without dialysis] and DON values [with dialysis pretreatment] for three rivers systems in France

River Name	DOC (mgC/L)	TDN (mgN/L)	DIN (mg N/L)	DON (mgN/L)	DOC/ DON
Vienne River					
Upstream (Eymouthiers)	2.74 (±0.05)	1.03 (±0.03)	0.87 (±0.01)	0.18 (± 0.03)	14.9
Upstream (Civaux)	3.66 (±0.04)	1.94 (±0.10)	1.75 (±0.02)	0.27 (± 0.04)	13.2
Downstream (Candes)	3.55 (±0.06)	3.26 (±0.05)	3.15 (±0.02)	0.19 (± 0.03)	19.0
Sèvre Niortaise River					
Upstream of abattoir	0.78 (±0.01)	10.37 (±0.18)	11.21 (±0.01)	0.33 (±0.05)	2.4
Downstream of abattoir	0.85 (±0.07)	9.82 (±0.08)	10.58 (±0.01)	0.34 (±0.07)	2.5
Upstream of Saint Maixent	0.94 (±0.01)	9.24 (±0.06	9.95 (±0.01)	0.28 (±0.07)	3.3
Downstream of Saint Maixent	0.98 (±0.01)	9.06 (±0.11)	9.86 (±0.01)	0.27 (±0.08)	3.6
Marne River					
Annet sur Marne (april)	2.29 (±0.06)	4.85 (±0.06)	4.60 (±0.01)	0.33 (±0.01)	6.9
Annet sur Marne (june)	2.02 (±0.07)	3.90 (±0.08)	3.54 (±0.01)	0.57 (±0.01)	3.5
Downstream of Meaux (june)	2.47 (±0.04)	3.24 (±0.01)	2.73 (±0.01	0.51 (±0.02)	4.8
Above Meaux (june)	2.14 (±0.04)	3.50 (±0.01)	2.90 (±0.01)	0.54 (±0.01)	4.0

Table 4.6 Data from the Cebron Reservoir basin [May 2004]

N°	Samples	DOC (mgC/L)	DIN (mgN/L)	DON (mg N/L)	DOC/DON (mgC/mgN)
1	Bodin swamp (downstream)	17.98	0	1.09 (±0.01)	16.5
2	Bodin swamp (upstream)	17.45	0.029	1.12 (±0.03)	15.6
3	Taconnière creek	12.41	2.381	0.74 (±0.06)	16.8
4	Raconnière creek	7.25	1.603	0.49 (±0.08)	14.8
5	Cébron creek	9.56	1.330	0.68 (±0.05)	14.0
6	Pool in Les couardes	23.63	0.199	1.75 (±0.05)	13.5
7	Ditch in Les champs Moines	15.80	2.505	0.90 (±0.04)	17.5
8	Ditch in Le Pressoux	21.42	0.630	1.37 (±0.02)	15.6
9	Pool in Les champs de la ligne	20.81	0.141	1.52 (±0.07)	13.7
10	Creek in Viennay village (downstream WWTP)	11.59	0.948	1.44 (±0.06)	8.0
11	Creek in Viennay village (upstream WWTP)	13.43	0.315	1.09 (±0.04)	12.3
12	Pool in Theil	10.41	0.075	0.82 (±0.07)	12.7
13	Cébron creek in Adilly village	8.86	1.700	0.65 (±0.14)	13.6
14	Cébron creek at Moulin Sunay	9.80	0.531	0.71 (±0.09)	13.8
15	La Lime creek	7.35	2.031	0.48 (±0.08)	15.3
16	Le Jussay creek	7.78	2.387	0.58 (±0.08)	13.4
17	Cébron reservoir "beach"	11.82	2.547	0.56 (±0.07)	21.1
18	Cébron reservoir "dam"	11.57	2.850	0.57 (±0.03)	20.3

Table 4.7 Water quality determined for a WTP on the Marne River, France

	pH	DOC (mgC/L)	TDN (mgN/L)	DIN (mgN/L)	DON (mgN/L)
Raw water	8.4	2.29	4.85	4.60	0.33 ± 0.01
Clarified water	7.5	1.42	4.89	4.58	0.53 ± 0.01
Filtered water (sand/anthracite)	7.5	1.37	4.76	4.49	0.45 ± 0.01
Ozonated water	7.7	1.54	4.85	4.56	0.23 ± 0.02
Treated water in reservoir	7.4	2.22	4.77	4.46	0.27 ± 0.02

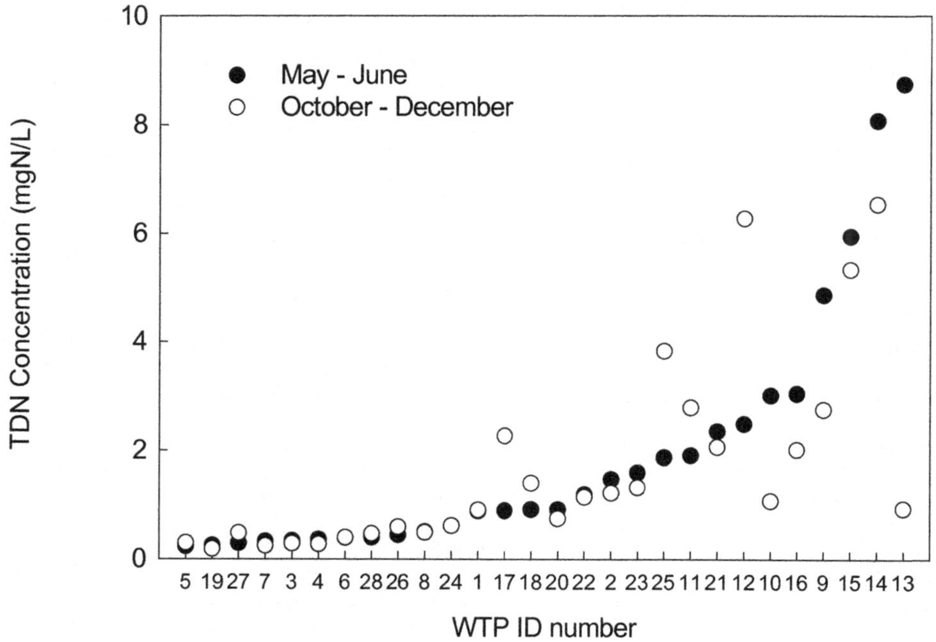

Figure 4.1 Distribution of TDN concentrations of raw waters from the full-scale WTP surveys. WTP ID number based upon facilities listed in Table 2.1.

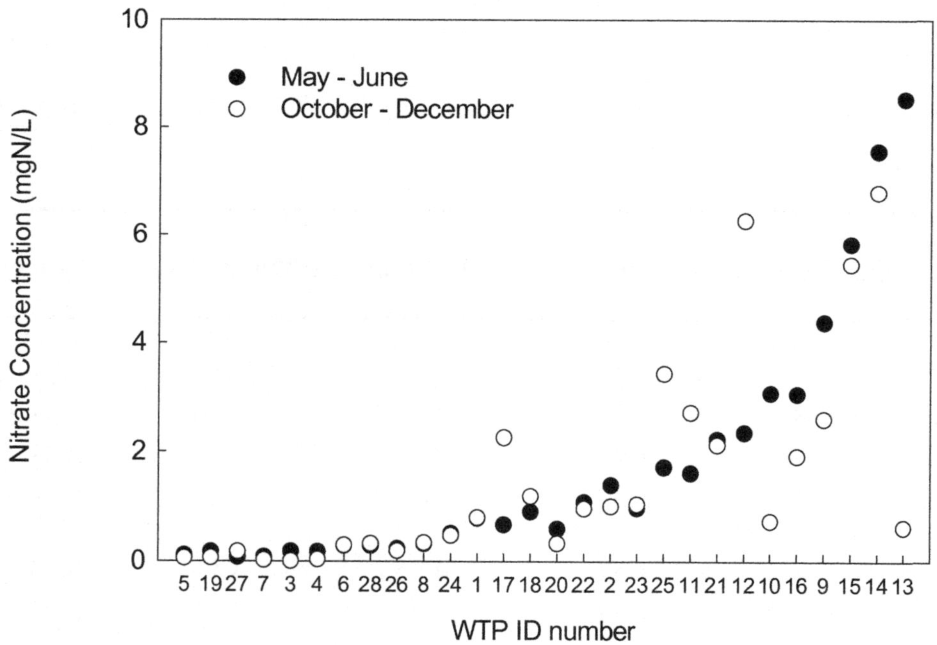

Figure 4.2 Nitrate concentrations of raw waters from the full-scale WTP surveys.

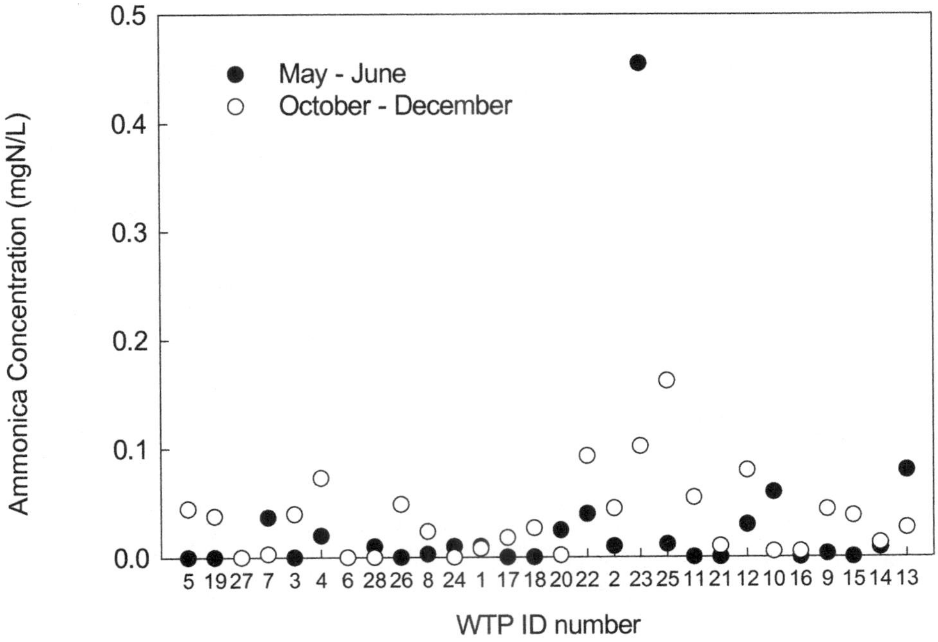

Figure 4.3 Ammonia concentrations of raw waters from the full-scale WTP surveys.

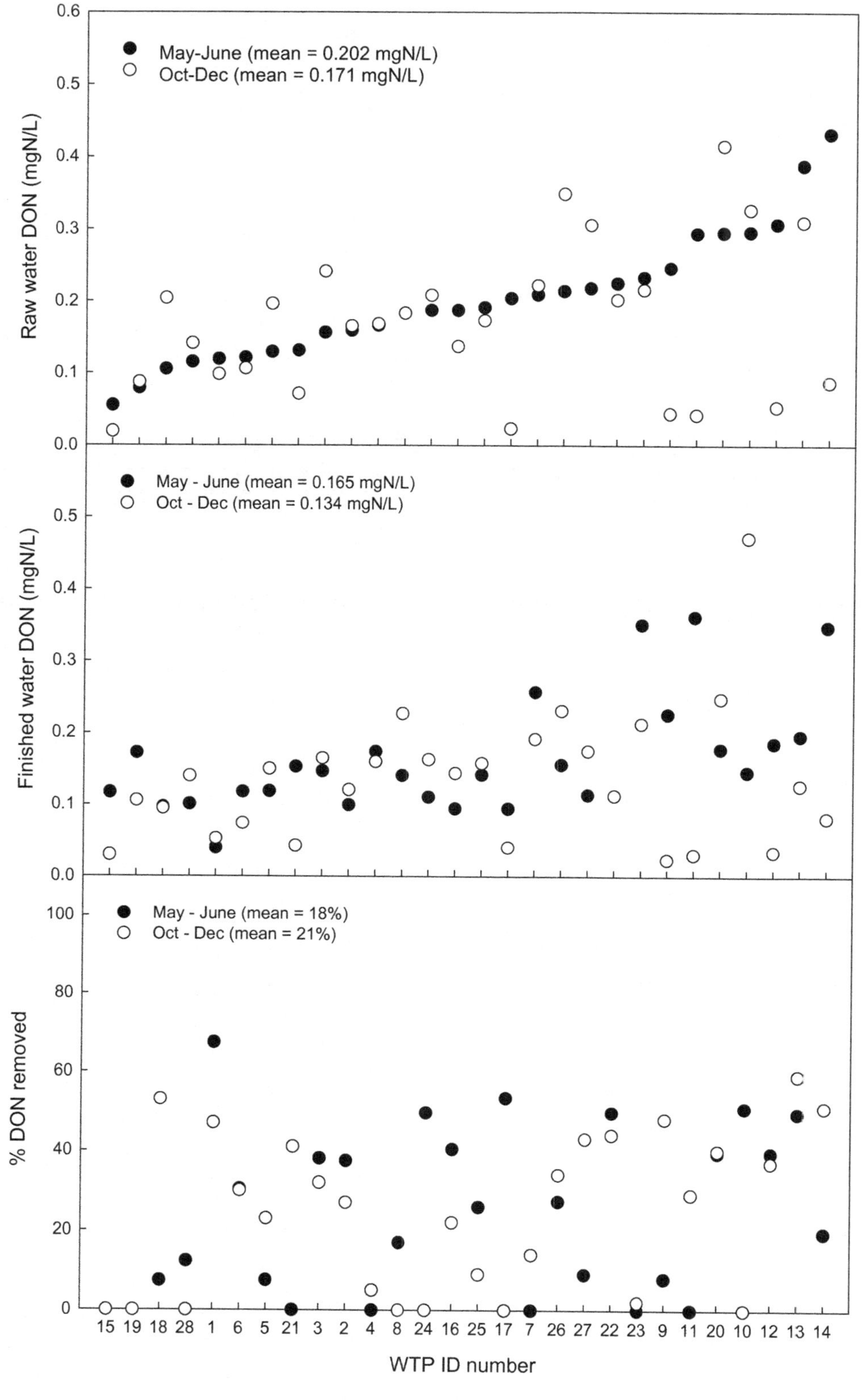

Figure 4.4 DON concentrations of raw waters and finished waters from the full-scale WTP surveys

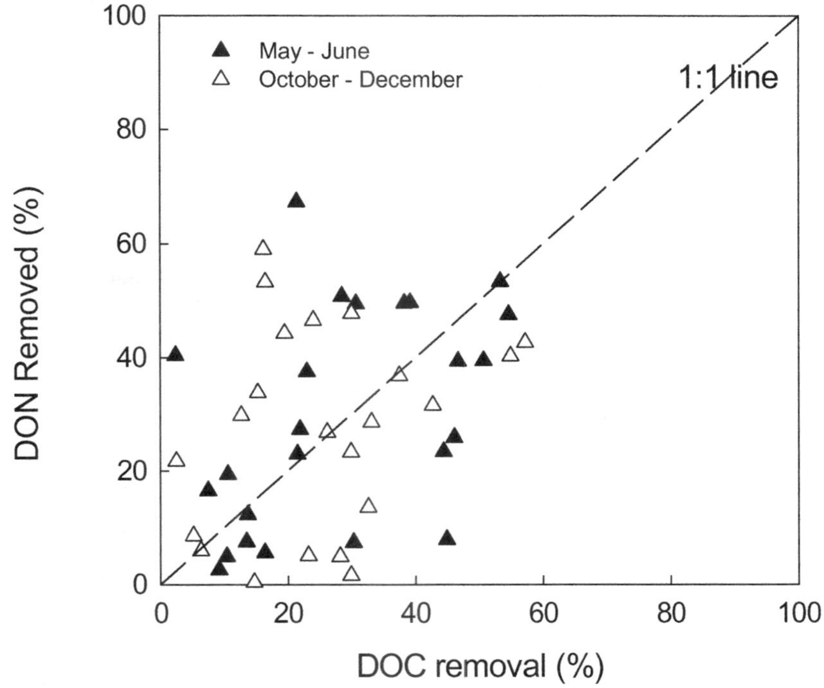

Figure 4.5 DOC removal versus DON removal from the full-scale WTP surveys

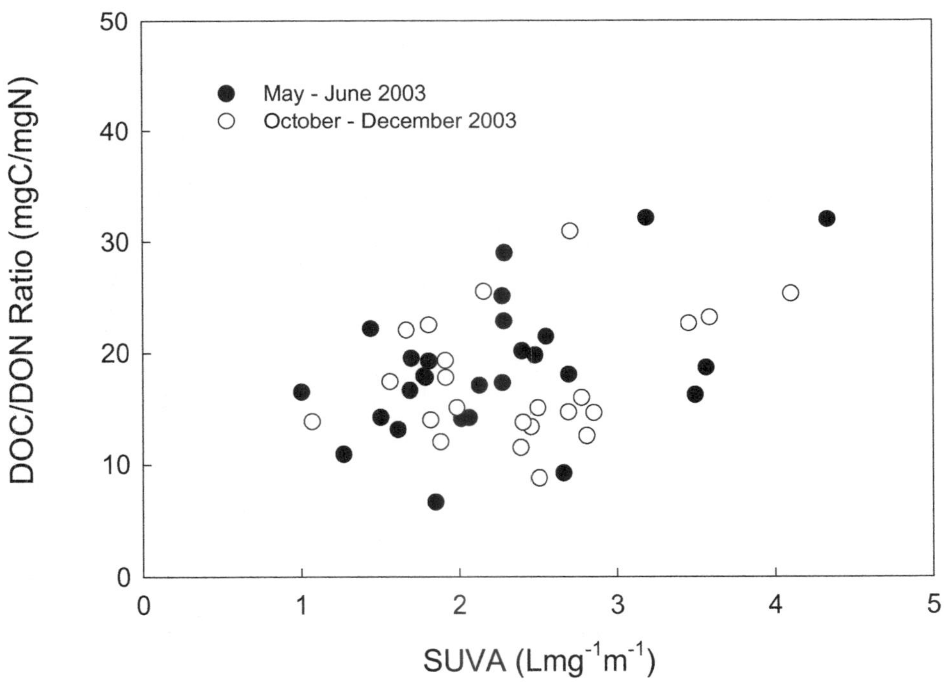

Figure 4.6 Plot of SUVA value vs. DOC/DON ratio of raw waters

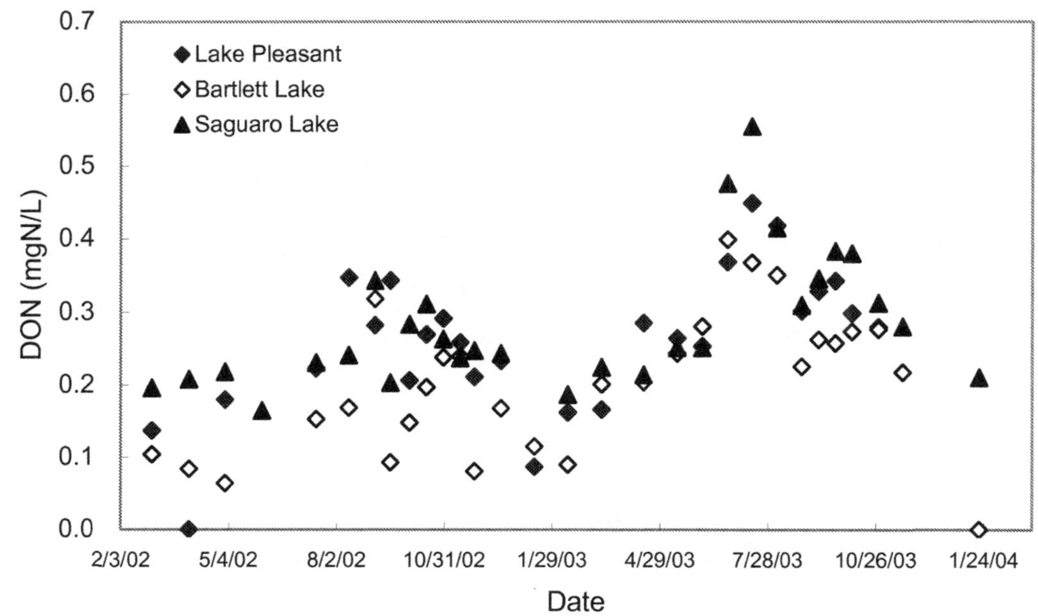

Figure 4.7 Seasonal variation of DON concentrations from three lakes in Arizona

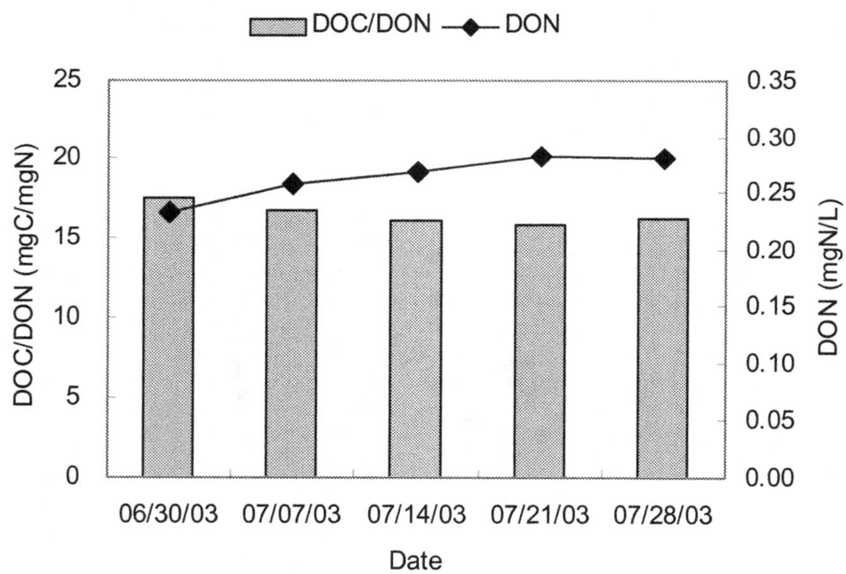

Figure 4.8 Temporal DON concentrations and DOC/DON ratios of El Capitan Reservoir water, California

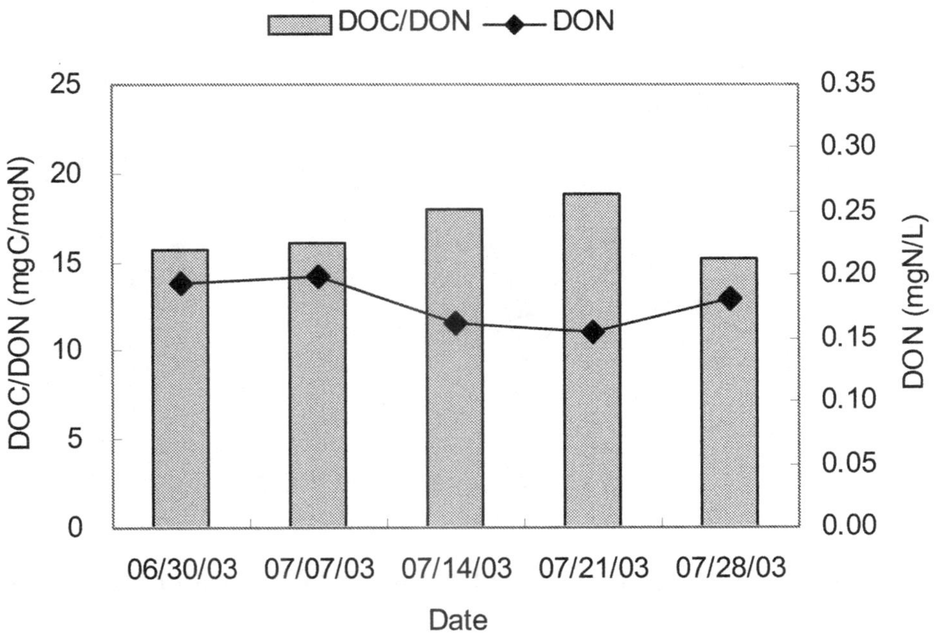

Figure 4.9 Temporal DON concentrations and DOC/DON ratios of blend of Colorado River and California State Project water, California

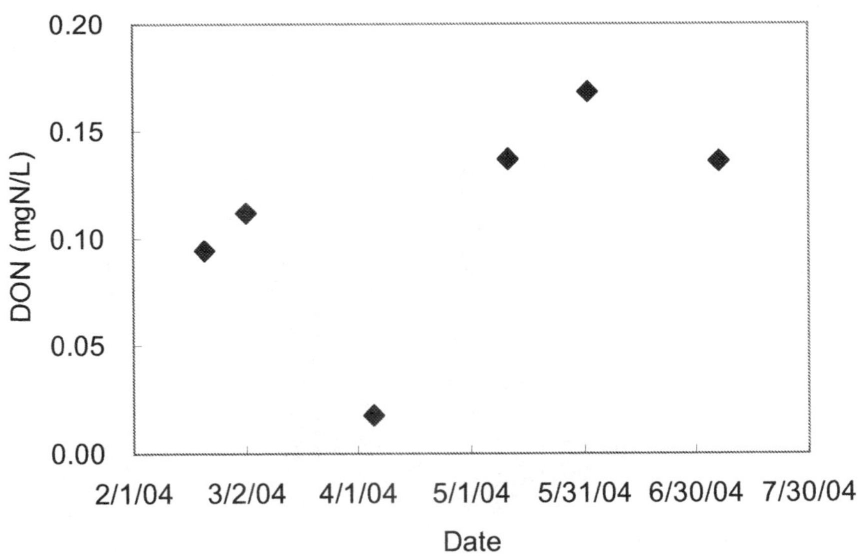

Figure 4.10 DON concentrations of a temporal survey from Muncie WTP, Indiana

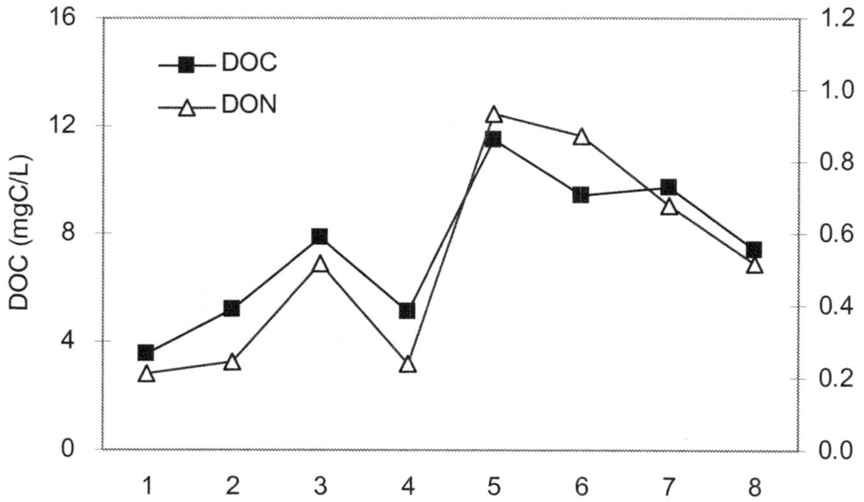

Figure 4.11 DOC and DON along the South Platte River [Wastewater effluent entered between points 2-3 and 4-5]

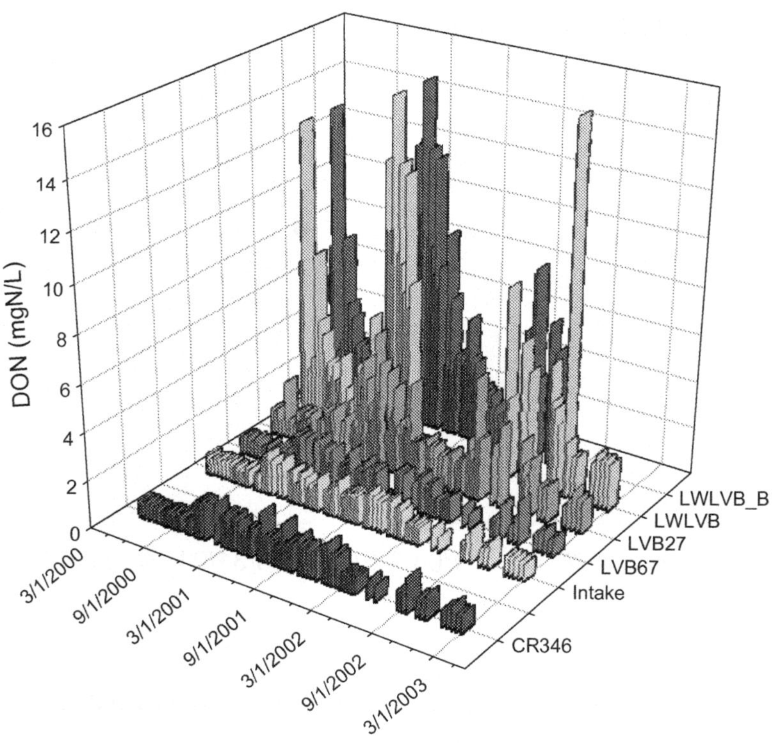

Source: Data provided by Southern Nevada Water Authority

Figure 4.12 DON concentration at a depth of 1 m below the water surface in Lake Mead, Nevada

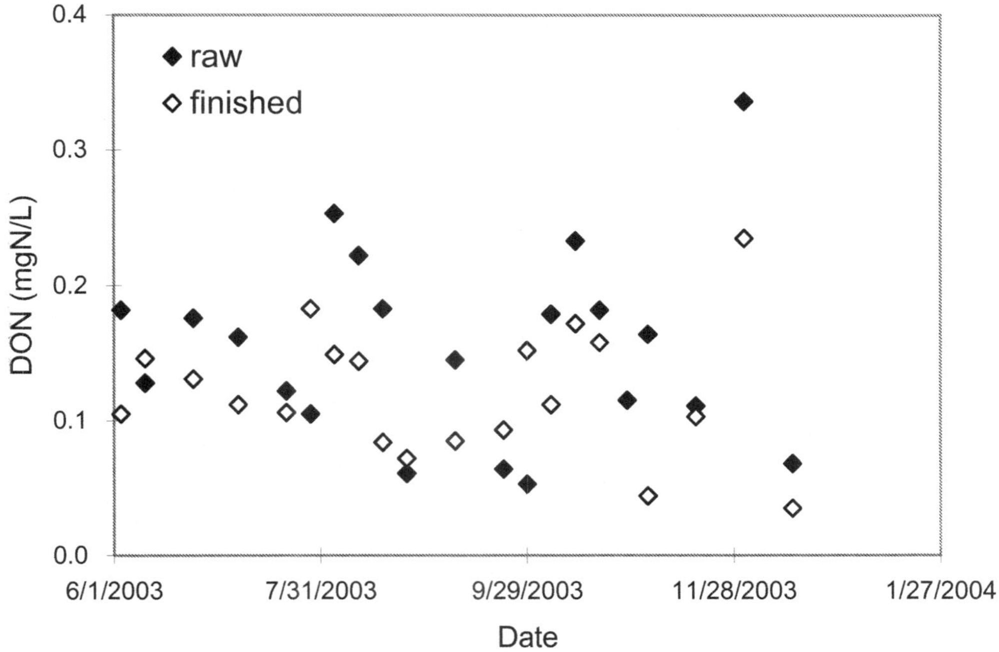

Figure 4.13 Seasonal variation of DON concentrations of raw and finished water from Green Way WTP, Arizona

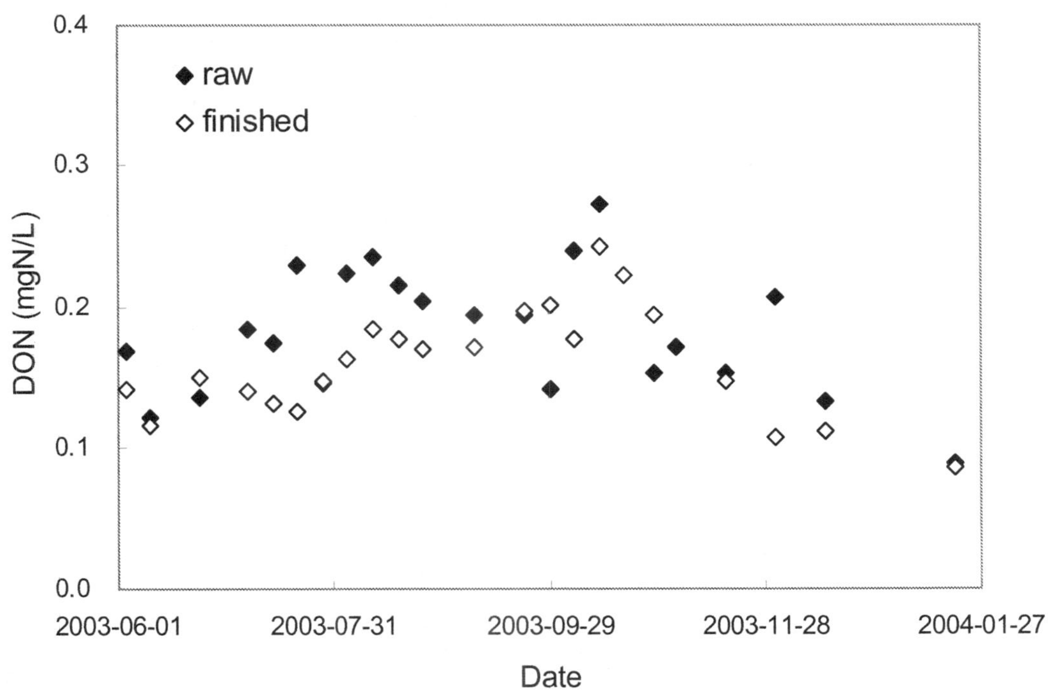

Figure 4.14 Seasonal variation of DON concentrations of raw and finished water from Union Hill WTP, Arizona

Figure 4.15 Map of France showing three river systems sampled

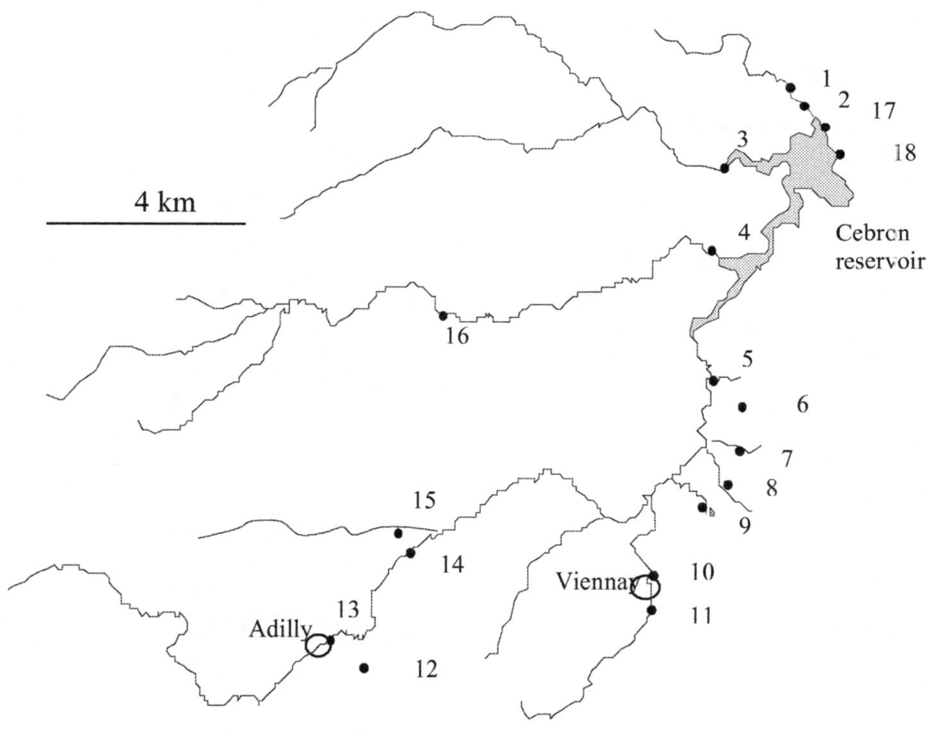

Figure 4.16 Map of the Cébron reservoir basin

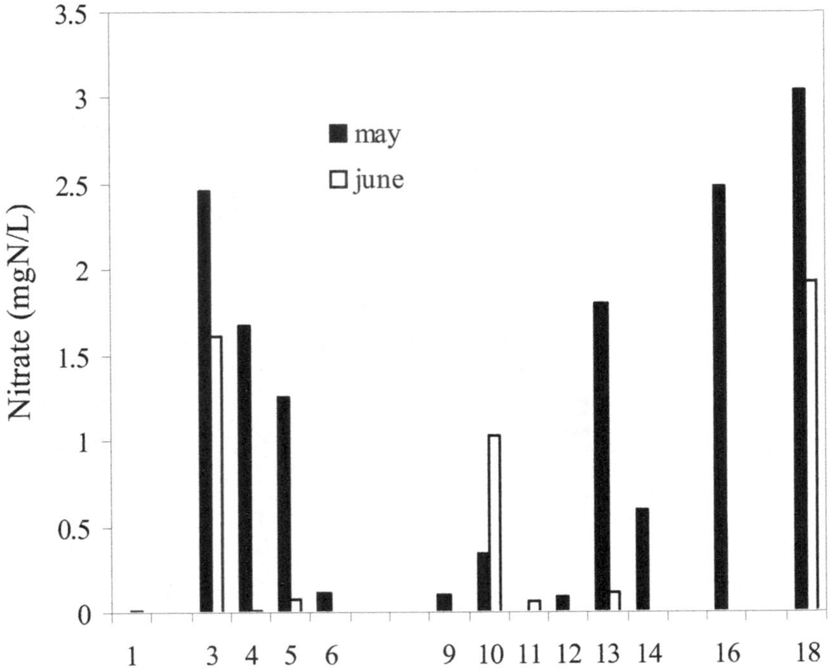

Figure 4.17 Nitrate concentrations from Cébron reservoir basin in 2004

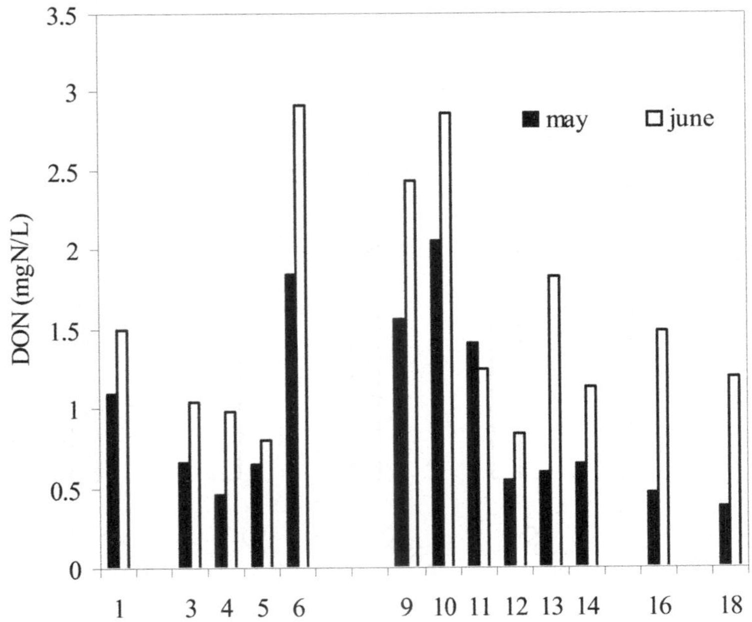

Figure 4.18 DON concentrations from Cébron reservoir basin in 2004

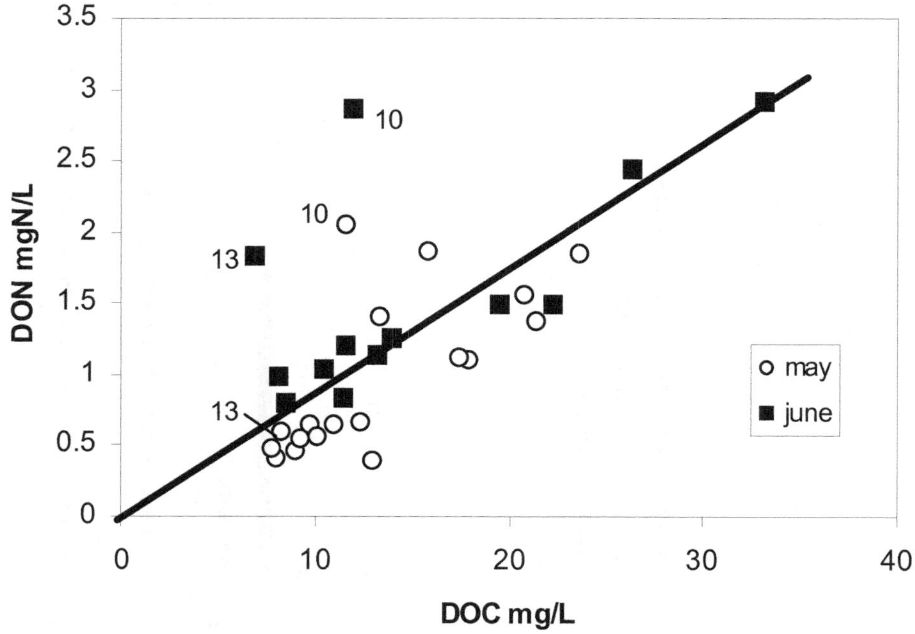

Figure 4.19 Correlation between DOC and DON from the Cébron reservoir basin in 2004

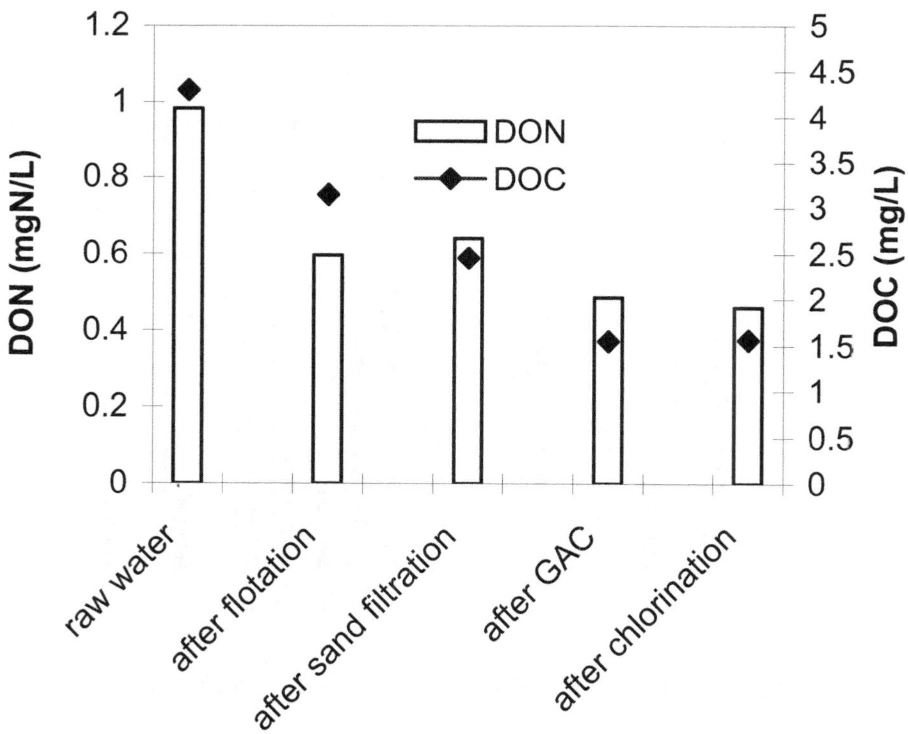

Figure 4.20 DON and DOC concentrations at the Saint Maixent WTP

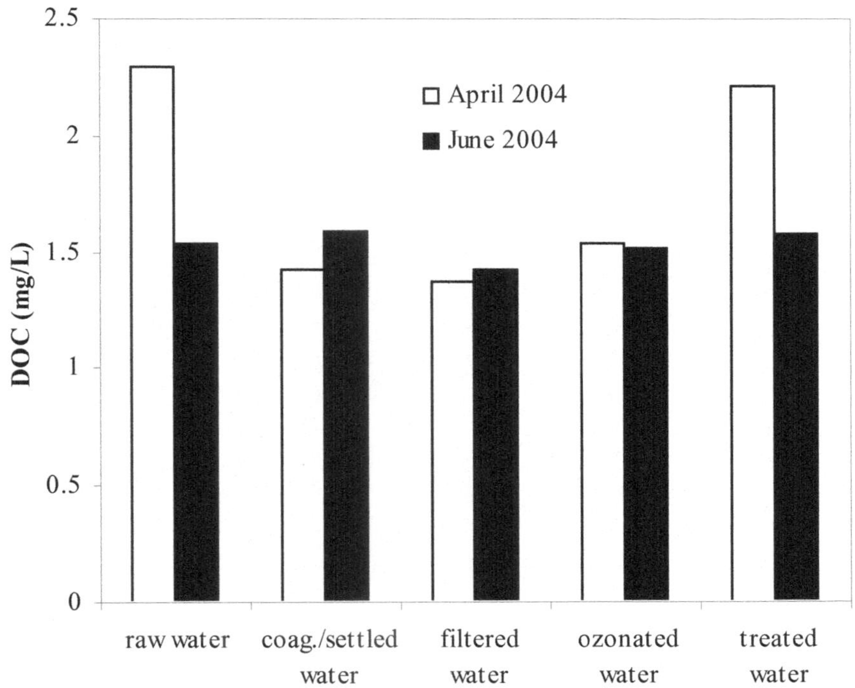

Figure 4.21 DOC concentration across a WTP on the Marne River, France

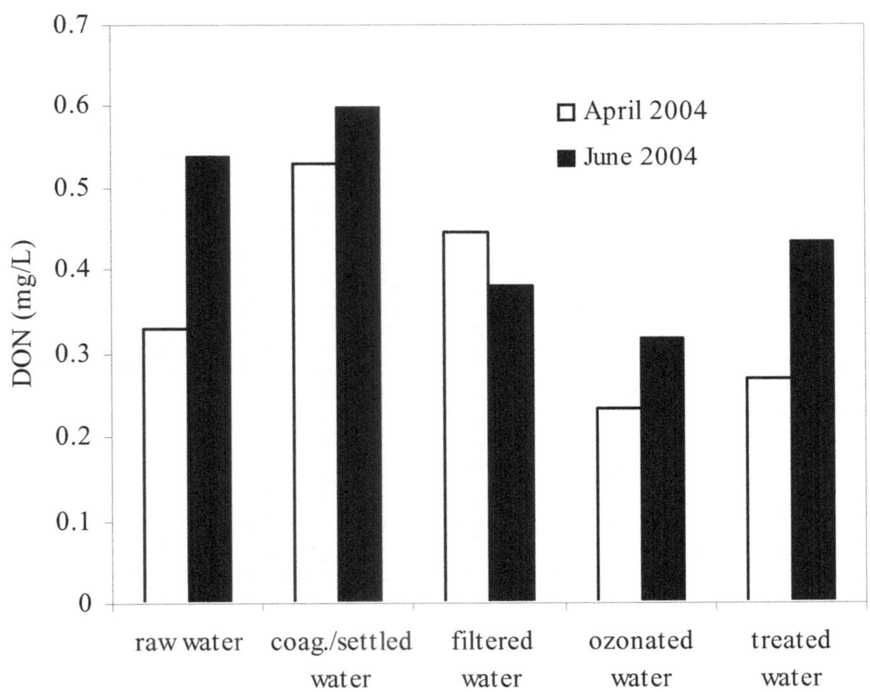

Figure 4.22 DON concentration across a WTP on the Marne River, France

CHAPTER 5 DISSOLVED ORGANIC NITROGEN CHARACTERIZATION

This chapter summarizes numerous approaches undertaken to characterize the size and structure of DON. In most cases the methods are applied to samples from drinking water treatment facilities to demonstrate their appropriateness, or in some cases inappropriateness.

MOLECULAR WEIGHT

During development of the dialysis pretreatment method (Chapter 3) it was concluded that less than 2% of the DON in drinking water samples permeated across the dialysis membrane. Only some free amino acids, or compounds such as urea, would have molecular weights of less than 100 dalton. Therefore, since the dialysis membrane has a nominal molecular weight cutoff of 100 dalton, it is reasonable to conclude that nearly all the DON in most drinking water samples have molecular weights of greater than 100 dalton.

Ultrafiltration was used to assess DON molecular weight. Samples from 16 raw and finished drinking waters collected during the second utility sampling campaign (October-December 2003) were used (see Chapter 4). Figures 5.1 and 5.2 present DON molecular weight distributions for raw and finished drinking waters, respectively. There was some variability in molecular weight distributions across the different utilities, but some DON was characterized into each of the three molecular weight ranges considered: < 1000 dalton, 1000 to 10,000 dalton, > 10,000 dalton.

Table 5.1 summarizes the results for raw and finished waters, and compares DON molecular weight distributions to those for DOC and UVA_{254}. In raw waters approximately one-third of the DON is in each of the three molecular weight ranges. There was slightly more DON in the <1000 dalton fraction than either DOC or UVA_{254}. In finished water there is a shift towards lower molecular weight DON, with the <1000 dalton accounting for approximately half the DON. Water treatment (primarily coagulation) preferentially removed higher molecular weight material, resulting the greatest shift in molecular weight distributions for UVA_{254} followed by lesser, but significant, changes in DOC and DON.

Molecular weight fraction of primary effluent and secondary effluent from a wastewater treatment plant was also quantified using ultrafiltration. Figures 5.3 and 5.4 summarizes the observations. Primary effluent is skewed towards lower molecular weight DOC and DON, with some higher molecular weight fractions. However, after secondary treatment the higher molecular weight fractions are not observed. It is possible that some colloidal organic matter was present in either sample, and could have adsorbed onto the membrane surfaces. However, mass balance balances on both DOC and DON were within 15% of the original mass applied to the ultrafiltration vessels, implying that significant losses of organic matter were not encountered.

ORGANIC CARBON TO NITROGEN RATIOS (NOM ISOLATES)

Table 5.2 shows for all studied NOM fractions the C/N ratios calculated from elemental analysis and C/N ratios determined from TN and TOC analysis. Glycin was used as a standard for nitrogen calibration. Hydrophilic acid fractions were more nitrogen enriched than hydrophobic acid fractions. The most polar fractions (HPI) commonly were usually the most nitrogen enriched. Colloidal fractions were also enriched with nitrogen compared against the hydrophobic acid fractions. Figure 5.5 plots for all fractions in Table 5.2. The elemental analysis and TOC/TDN analysis give similar values of C/N ratios, and thus aided in verifying the ability of high temperature combustion methods to completely oxidize nitrogenous materials.

AMINO ACID AND AMINOSUGAR CONTENT OF NOM ISOLATES

NOM Isolates Characterization

Table 5.3 shows the total dissolved amino acids concentrations of 28 fractions analyzed using HCl hydrolysis and AccQTag method. Figures 5.6, 5.7, and 5.8 show the distribution of the different classes of amino acids including aminosugars based on nitrogen content for all fractions studied. Neutral amino acids represent the major group of amino acids in NOM isolates as already discussed in the literature. As general observations:

- The TPI fraction is more enriched in neutral aminoacids and aminosugars as compared to its corresponding HPO fraction
- The HPO fraction is more enriched in acidic aminoacids as compared to its corresponding TPI fraction (with the exception of Gartempe NOM isolates)
- The HPO fraction is more enriched in aromatic aminoacids as compared to its corresponding TPI fraction, observation in accordance with SUVA. However, for EfOM Naintré isolate we do not confirm the high proportion of aromatic amino acids observed for Saint Julien EfOM isolate. This difference may be due to the fact that Saint Julien is a low-load WWTP whereas Naintré is a high-load WWTP.
- Colloid fractions from EfOM or NOM are enriched in aminosugars compared to HPO and TPI fractions.

Martens and Loeffelmann showed that methanesulfonic acid (MSA) hydrolysis coupled with nonderivatized amino acids and aminosugars quantification by ion chromatography with pulsed amperometric detection increased the total recovery of organic N as amino acids (AAs) and aminosugars (ASs) in soil organic matter (Martens and Loeffelmann 2003). Therefore, it was decided to test the MSA hydrolysis procedure coupled with the AccQTag derivation method. The MSA hydrolysis procedure from Martens and Loeffelmann (2003) was modified slightly as follows:

- A sample volume of 100 µL was evaporated using the PicoTag Waters station in a hydrolysis tube,
- 30 µL of 4N MSA solution were added and hydrolysis tubes were heated at 165°C for 1 hour under N_2 atmosphere.

After one-hour hydrolysis, MSA was neutralized with 35 µL of 4N NaOH and total volume was evaporated under vacuum pressure (termed Derivation with the AccQTag method.). Derivation reagents were added and samples injected for chromatography quantification. Chromatograms showed no amino acid peaks indicating that in these conditions MSA hydrolysis procedure are probably not compatible with AccQTag derivation method. The OPA method was instead used to test the MSA hydrolysis procedure.

Two EfOM isolates (Saint Julien TPI and Saint Julien Colloid) were hydrolyzed with liquid phase MSA protocol (termed Derivation with the OPA method). Pre-column derivation of amino acids with orthophtaldialdehyde (OPA) was performed using Merck AS 4000 auto sampler. Amino acids were chromatographed on a Delta Pak WATERS C18 column (5μm, 100 Å, 3,9x150 mm) and with a gradient elution of 2 eluants (acetate/monohydrogenophosphate buffer 50mM pH 7.4; methanol/water 75/25). Tetrahydrofuran was added to the acetate/phosphate buffer to improve the separation between glycin and threonin. With this detection method, aminosugars are not analyzed.

Tables 5.4 and 5.5 show amino acids concentrations determined after HCl hydrolysis (AccQTag detection method) and after MSA hydrolysis (OPA detection method). Analysis was done on 5 replicates. For Saint Julien colloid and TPI fractions results obtained with the MSA-OPA method are significantly higher than the results obtained with the HCl-AccQTag method. More work needs to be done to confirm that these results are related to the better efficiency of the MSA hydrolysis.

Derivatization of the amine functional groups using AQC

The objective of this part was to examine the possibility to use the AQC derivatization agent (derivatization agent used for total dissolved amino acid and aminosugars analysis) for the analysis of amino structures using HPSEC coupled with fluorescence detection. The derivatization protocol was as follow: 90 μL of NOM solutions (20 mg/L of NOM) were added to 90 μL of borate buffer and 60 μL AQC. For the MQ + AQC sample, the NOM solution was replaced by the same volume of MilliQ water. For the NOM solution controls (without derivatization) 90 μL of NOM solution (20 mg/L of NOM) were added to 90 μL of borate buffer. The HPSEC analysis was operated using Protein-Pak Diol (OH) 10μm (7.8 x 300 mm) SEC column. The mobile phase prepared was made of 20mM de NaH_2PO_4 and 0.1 M of NaCl. The wavelength of the UV detector was set a 260 nm. For fluorescence measurement, the excitation wavelength was 240 nm and the emission wavelength 395 nm (analytical condition for the AMQ detection produced from the hydrolysis of AQC).

Figure 5.9 shows the HPSEC chromatograms obtained for the NOM fractions isolated from the Loire River (LE) and Britany River. The HPSEC chromatograms obtained after derivatization do not show any specific chromatographic band that could be related to derivatized amine functional groups. After derivatization with AQC the chromatograms recorded correspond to the superposition of the AMQ chromatogram (compound formed from the hydrolysis of AQC) and the chromatogram of the NOM solution without addition of the derivatization agent (better observed with HPO that exert a stronger fluorescence signal due to its stronger aromatic character).

HPSEC ANALYSIS FOR NITROGENOUS ORGANIC MATTER

Solutions of NOM (20 mg/L) were analyzed by HPSEC coupled with UV and fluorescence detection. The UV absorbance wavelength was set at 260 nm and the fluorescence excitation and emission wavelengths were 280 and 320 nm, respectively. Coble and others mentioned that this particular fluorescence domain may correspond to protein-like (tyrosine, tryptophane) and phenol-like fluorophores (Chen et al. 2003; Coble 1996). Figure 5.10 gives as an example the normalized (in intensity) HPSEC chromatograms of the colloid, HPO and TPI fractions isolated from a river in Britany (France). Model compounds (BSA protein and tryptophane amino acid) were also analyzed and their HPSEC/fluorescence profiles are included for comparison.

The HPSEC/UV chromatograms are typical of NOM SEC chromatograms published in the literature. The colloid chromatograms (both UV and fluorescence detection) exhibits a low intensity band with high molecular weight, higher than the 70,000 dalton MW of BSA protein, which can be related to protein-like biopolymers. The TPI NOM chromatogram is shifted toward higher retention times as compared to HPO and colloid NOM, observation that relates to its lower average molecular weight. The HPSEC/fluorescence chromatograms give a thinner band (lower intensity as compared to the UV signal, not shown) after 11 minutes that correspond to low molecular weight organic structures. However, the comparison with the tryptophane band indicates that these structures have higher MW than free amino acids.

Plots were developed for the area of the HPSEC/fluorescence band as a function of the aromatic amino acids content of some NOM fractions. Tryptophane cannot be analyzed by the AccQTag method, consequently, the aromatic amino acids content corresponds to the sum of tyrosine and phenylalanine only. No correlation could be drawn with this set of data, which would indicate that this fluorescence band is more related to phenol fluorophores than peptide-like fluorophores.

ULTRAVIOLET RATIO INDEX (URI)

HPLC with UV detection was used to demonstrate the URI method. The ratio of ultraviolet absorption at 210 nm to 254 nm after HPLC has been suggested to yield information regarding nitrogenous organic material (Her et al. 2004). Therefore it was applied to a set of samples from the South Platte River system which passes through metro-Denver and is impacted by wastewater discharges (reference Figure 2.4 for sampling sites). Measured parameters for the sites are summarized in Table 5.6.

The URI values were determined for four peaks in each chromatogram. The first peak is in the protein/polysaccharide region (eluting at about 2000 seconds), the second and third peaks are in the humic substance region (eluting at about 3000 and 3500 seconds, respectively), and the fourth peak is in the low-MW acids region (eluting at about 4000 seconds). An example of URI values determined corresponding to a chromatogram is displayed in Figure 5.11. The summaries of URI values for all the peaks in all the chromatograms (Table 5.7) show that the URI values for the first and fourth peaks are generally greater than those of the second and third peaks. This implies that the first and fourth peaks are more representative of substances containing DON (e.g., proteins and amino acids), or that peaks found in the protein/polysaccharide region and the low-MW acid region are more representative of substances containing DON. For simplicity, the peaks found in these two regions will be referred to as the nitrogen-containing peaks, or the N-peaks. The N-peaks can be compared between samples to help describe DON content at different points in the study segment.

ULTRAVIOLET ABSORBANCE (WITHOUT HPLC)

It was hypothesized that spectroscopic measurements may be surrogates for DON or DOC/DON ratios. DOC/DON ratios for raw waters from the utility survey (Chapter 4) exhibited a weak correlation (r=0.40) with SUVA. However, a general trend was present for increasing SUVA and higher DOC/DON ratios. For 14 NOM isolates DOC/DON ratios correlated (r=0.82) with SUVA. However, colloidal NOM isolates had low DOC/DON ratios and high SUVA and fell well below the line representing the other NOM isolates. Thus, the raw drinking water samples may have contained a mixture of NOM isolates, including colloidal materials.

Ultraviolet absorption at 210 nm may be indicative of protein-like organic material (Her et al. 2004). For raw water samples from the utility survey nitrate correlated ($r>0.90$) with UV absorbance at wavelengths between 200 and 225 nm. Nitrate standards prepared in Nanopure water yielded an extinction coefficient of 7490 $M^{-1}cm^{-1}$ ($r=1.0$). For the raw water sample, the UVA_{210} absorbance was not correlated with DON. To account for nitrate, the observed UVA_{210} was corrected for nitrate content using the extinction coefficient and measured nitrate concentrations. After subtracting the portion of UVA_{210} attributed to nitrate in the samples, the corrected UVA_{210} exhibited a weak correlation ($r=0.47$) with measured DON. Other inorganic ions potential absorb at 210 nm. To eliminate potential inorganic interferences experiments were conducted using NOM isolates dissolved in Nanopure water. A weak correlation ($r=0.40$) was observed between UVA_{210} and DOC/DON for 20 NOM isolates. A better correlation ($r=0.60$) was observed between DOC/DON and the ratio of UV absorbance at 210 nm to 254 nm. Therefore, on bulk water samples the ratio of UV absorbance at 210 nm, corrected for the presence of nitrate, to 254 nm may be a reasonable indicator for the DOC/DON ratio. The use of HPLC to separate out potentially interfering ions probably improves the use of this ratio.

EXCITATION-EMISSION FLUROESCENCE SPECTROSCOPY

Previously methods have been developed to characterize NOM isolates and wastewaters using fluorescence (Chen et al. 2003; Her et al. 2004; Westerhoff et al. 2001). Specific regions of excitation-emission matrix (EEM) fluorescence are indicative of protein-like substances, while other regions are more indicative of humified materials. For raw water samples, three EEM regions were correlated with DON or DOC/DON ratios. DON correlated ($r=0.71$) with the protein-like EEM region (excitation 270-280 nm and emission 300-350 nm). DOC/DON ratios correlated ($r=0.6$) with the ratio of either the fulvic-like (excitation 215-250 nm and emission 388-431 nm) or humic-like (excitation 320-335 nm and emission 405-440 nm) regions divided by the protein-like region. DOC correlated with both the fulvic-like ($r=0.89$) and humic-like (0.93) EEM regions.

EEMs can be mathematically manipulated to aid in analysis of these complex datasets that become developed. One example of differential EEM fluorescence will be discussed for a sampling location on the South Platte River that is impacted by upstream wastewater discharges during low-flow events, but has less of an influence under high-flow (runoff) conditions. Subtraction of the high-flow day from the low-flow revealed a peak maximum in the protein-like substance region (Figure 5.12).

EEMs for three samples from the DON field survey in the USA are shown in Figure 5.13. The shapes of the fluorescence intensity contours can be read like a topographic map. The protein-like region indicated in Figure 5.12 is present in all three samples, but not at the same intensity nor cause the same shape of the contours. In recent work, the Investigators took the approach of integrating the surface volumes below the contours as an indicator of protein-like fluorescence intensity (Chen et al. 2003). Apply that concept to the three EEMs in Figure 5.13 reveals "steeper topography" in the protein-like region for the lowest DON sample compared against the "topography" for the higher DON samples.

PYROLYSIS-GC/MS

Pyrolysis-GC-MS requires pretreatment to remove salts and concentration via lyphilization. For this study salt was removed using XAD-8 chromatography. After lyphilization it was determined that 30% to 70% of the initial carbon and nitrogen was lost during these pretreatment steps. Table 5.8 summarizes pyrolysis GC/MS results for samples collected during high-flow on the South Platte River. Large percentages of material were non-characterized by pyrolysis GC/MS. Aminosugars and polysaccharide fractions increased at sites influenced by wastewater.

PROTEIN ASSAY TECHNIQUES

Lowry assay, BCA assay, and Bradford assay have been used to measure protein contents in solution with high protein levels such as wastewater. However, the Lowry assay is very sensitive to interference with detergents, carbohydrates or reducing agents. The BCA assay is similar to the Lowry assay and also sensitive to interferences with reducing agents. The Bradford assay has high protein-to-protein signal variability and is not compatible with detergents. Upon BCA and Bradford methods, samples must be read in a short time (e.g., BCA method within 10 minutes). Furthermore, those three methods can measure high protein levels ranging 0.5 mg/L to 1500 mg/L; they are not suitable for surface water or drinking water.

NanoOrange protein assay (Molecular Probes Inc., OR) is a possible protein assay in solution with low protein levels. When bound to proteins in the diluents provided, the NanoOrange assay reagent has a broad excitation peak centered at about 470 nm and a broad emission peak centered at about 570 nm. The unique chemistry of the NanoOrange reagent allows for accurate detection of proteins in solution at concentrations between 10 μg/L and 10 mg/L. This level of sensitivity is considerably better than that achieved with the Lowry assay, the BCA assay, or the Branford assay. However, the NanoOrange assay is not as established as other assays. The followings are results of preliminary experiments using NanoOrange assay.

Protein content using NanoOrange was applied for 28 surface water samples using the NanoOrange assay and found that the assay overestimates protein contents. In most samples, protein concentrations were higher than DON concentration (protein concentration should be less than DON). It was hypothesized that high TDS interferes with the assay because samples with low conductivity (< 100 μS/cm) showed lower protein concentration than DON.

To verify the interference between ion concentration and the assay, a 3 mg/L BSA solution was prepared with different anions and cations added in concentrations 1 to 10 mM. Samples were prepared in triplicate and BSA concentrations of the samples were measured. Results exhibited significant interference by divalent ions (Table 5.9). In case calcium chloride, milky emulsion was observed over 1 mM. Therefore, it was concluded that the NanoOrange method would not be suitable for natural water samples. However, it might be valid for solutions prepared by dissolution of NOM isolates or application to natural water samples after dialysis pretreatment removed nearly all the inorganic ions.

Protein content was measured using NanoOrange for 25 NOM isolate solutions. NOM isolates were dissolved in deionized water (DOC range: 8 to 100 mg/L; DON range: 0.3 to 3.7 mgN/L) and had no interference from ions. The NanoOrange method is calibrated to bovine serum albumin (BSA); here it is reported as mgBSA-N/L by accounting for the nitrogen content of BSA. The average protein content for the 25 NOM isolates was 0.15 mgBSA-N/mgDON (range: 0.021 to 0.475 mgBSA-N/mgDON). Calibration curves indicated the NanoOrange MDL was 0.015 mgBSA-N/L, but could be lower by increasing the slit width on the fluorimeter. NOM (HPO, TPI, HPI) isolated from lake or reservoir waters presented higher proportion of protein than those isolated from river water or finished drinking water (Table 5.10). Colloids fraction exhibited higher protein contents possibly due to bacterial cell fragments in colloid fraction. The NanoOrange method has a much lower detection limit than the Lowry or Bradford methods, and has the potential to characterize nitrogenous organic materials in low ionic strength solutions.

^{15}N-NMR SPECTROSCOPY

Chemical shift values were obtained for the principal lines in ^{15}N NMR spectra of two solid soil extracts humic acid and fulvic acid, for all lines in the amino acids glycine and L-arginine, and for the nucleic acid uracil. The solid soil extracts show broad spectra, but each has a single large peak arising from similar residues in these extracts, as discussed below (Figure 5.14). Referenced to the primary standard nitromethane, the shift for the main humic acid peak is -262.8 ppm, while that of the main fulvic acid peak has the value -264.7 ppm. NMR signals from these extracts, which were not enriched or "labeled" with the isotope ^{15}N, were not strong, and as many as 168,000 acquisitions were obtained over periods of more than a week. Signal-to-noise ratios were only around 3 for 168,000 acquisitions in humic acid, and 2 for 20,000 acquisitions in fulvic acid.

The weakness of the signals in the soil-derived samples is partly to be expected from the breadth of their spectra (Kogel-Knabner 1997), which appeared to be in excess of 20 ppm total. Even the large peaks themselves had breadths of several ppm, a situation typical for soil extracts but one that results in large numbers of acquisition being required for acceptable signal-to-noise ratios. Acquisition times were also necessarily long, owing to the long pulse delays typically required between individual data acquisitions (10 seconds).

The observed ^{15}N peak shift values in the soil-derived samples analyzed herein may represent residues from very similar compounds. In general there are many N-containing compounds that fall within a given shift range, but one group that would not be unexpected in soil extracts is hydrogen-bonded oligopeptides with glycine residues. Kuroki et al. measured both ^{13}C and ^{15}N spectra in ten of these compounds, finding ^{15}N shifts from -269.45 to -258.05 ppm. Our values for the main peaks of the soil-derived samples fall in the middle of this range, leading us to tentatively attribute them to these residues. Kuroki et al. were interested to compare trends in their observed shifts to measured lengths of the O∙∙N and N-H linkages in the hydrogen bonds of the oligomers. They found no correlation with O∙∙N distance but a strong linear correlation with N-H distance.

Measured shift values for the solid amino acids are considerably more shielded than those of the soil-derived samples. The strong single line of glycine lies at -348.2 ppm, while the four observed peaks in L-arginine occurred at – 297.4, -305.9, -344.6 and -349.5 ppm (see spectrum below). The first two of these peaks probably arise from the guanidino nitrogens of L-arginine, as they are the least shielded (see structure of the L-arginine molecule, also below). The alpha-NH_2 peak should be the most shielded, at -349.5, leaving the internal peak at -344.6 ppm. The presence of four peaks for L-arginine indicates that the terminal guanidino nitrogens are not equivalent in the solid. In solution the number of observed peaks for L-arginine is three.

The amount of spectrometer time required make these tasks difficult to schedule on a busy NMR instrument, in an active research laboratory. To effectively implement a broad program of NMR measurements the best course of action would be to use one of the most modern, ultra high-field instruments, operating at a frequency for protons of 800 MHz or higher. While it is possible to obtain ^{15}N-NMR spectra, only selective nitrogen bonds can be evaluated and are similar to those detected by FTIR. Therefore ^{15}N-NMR spectra probably do not have a major role in advancing the understanding of DON structure with regards to water treatment applications.

SUMMARY

The following conclusions regarding characterization of nitrogenous organic materials in drinking waters were reached:

- The carbon to nitrogen ratio for NOM in drinking water ranges between 10 to 30 mgC/mgN. Transphilic and colloidal NOM isolate fractions are more nitrogen enriched (DOC/DON between 5 to 10 mgC/mgN) while hydrophobic acids generally have higher DOC/DON ratios.
- Based upon dialysis, >95% of the DON has molecular weights greater than 100 dalton.
- Based upon ultrafiltration separation, the molecular weight distribution of DOC and DON in drinking waters are comparable. In raw drinking water, DON was equally distributed (~ 1/3rd each) into the three ranges considered: < 1000 dalton, 1000 to 10,000 dalton, and >10,000 dalton. In finished drinking water, the distribution was skewed towards the lower molecular weight fractions, with approximately half of the DON < 1000 dalton and then equal amounts of the other two molecular weight ranges.
- Based upon high performance size exclusion chromatography (HPSEC) with DOC, UV, and fluorescence on-line detection indicated the presence of high molecular weight biopolymers for the colloidal NOM isolates. The data also confirms that most of the DOC has molecular weights greater than that of free amino acids. Analysis of the HPSEC with fluorescence detection data suggests the fluorescence is attributed to aromatic materials other than tryptophan or other nitrogenous aromatic structures.
- HPLC analysis with UV detection at multiple wavelengths (210nm to 254 nm ratio of absorbances) was used to the compute the URI value of NOM, and hypothesized that the URI value was indicative of proteinacous materials. Two HPLC peaks corresponding with protein/polysaccharides and low molecular weight materials exhibited high URI values, and considered to be representative of nitrogenous materials.
- Simple spectroscopy measurements (SUVA), with HPLC, also provided reasonable correlations to DON or DOC/DON ratios. After correcting UVA_{210} for the presence of nitrate, the ratio of corrected UVA_{210} to UVA_{254} correlated with DON. The protein-like intensity at specific fluorescence wavelengths also correlated with DON.
- Methods were applied for amino acid and aminosugar quantification. Using methanesulfonic acid for hydrolysis of NOM isolates yielded higher amino acid content compared to the use of hydrochloric acid for hydrolysis. For a colloidal NOM isolate MSA hydrolysis yielded an amino acid concentration equivalent to 584 μgN/L, or 53% of the DON, compared to only 316 μgN/L, or 30% of the DON, by HCl hydrolysis. Similar improvement in hydrolysis yields were observed for a TPI NOM isolate, where MSA yielded amino acids that accounted for 21% of the DON compared to only 8% after HCl hydrolysis.
- Different NOM isolates exhibited variable levels of amino acids. Glutamic acid, glycine, lysine and arginine were among the highest concentration in many samples. At the time only HCl hydrolysis was being employed. The colloidal fraction was most enriched with aminosugars. The HPO isolate was more enriched in acidic and aromatic amino acids

compared against a TPI fraction from the same water source. The TPI fraction was enriched in neutral amino acids and aminosugars.

- Pyrolysis GC-MS analysis required desalting of bulk waters, which can result in significant loss of NOM. Future work should consider the use of dialysis as a desalting step rather than resin extraction. The protein, peptides, and aminosugar fractions determined by pyrolysis GC-MS accounted for approximately 3% to 5% of the signal. This is in reasonable agreement with nitrogen content of NOM.
- ^{15}N-NMR was investigated as a potential means to augment standard FTIR analysis. However, due to the low nitrogen isotopic content of NOM very long NMR run times are required to acquire meaningful signals, although it was quite easy for model nitrogen compounds. After acquiring ^{15}N-NMR spectra for a IHSS soil humic acid spectra peaks indicative of amine bonds were observed. Overall, ^{15}N-NMR may hold promise for nitrogenous NOM characterization but for the amount of effort required to obtain sufficient NOM mass for analysis, FTIR analysis could yield comparable information with less than 1/10th the mass of material.

Several commercially available protein test-kits were evaluated. The Lowry and Bradford assays are appropriate for untreated or poorly treated wastewaters, but are not sensitive enough for low protein concentrations such as found in reclaimed wastewaters or drinking waters. A biomedical assay (NanoOrange) used to measure protein content in cells is very sensitive, but exhibited interferences from salts in water, especially from divalent ions. As such it is of limited use for bulk drinking water samples, but could be used on samples after dialysis pretreatment. For NOM isolates, the NanoOrange method could be used without concern for salts. The colloidal fraction exhibited the highest protein content based upon the NanoOrange reagent.

Table 5.1 Averaged MW distributions for DON, DOC, and UVA_{254} in raw and finished samples from 16 WTPs

Sample and Molecular weight range (Dalton)	DON	DOC	UVA_{254}
Raw Water			
<1000	37%	32%	26%
1000 to 10,000	30%	36%	38%
>10,000	33%	32%	36%
Finished Water			
<1000	47%	43%	57%
1000 to 10,000	29%	34%	29%
>10,000	24%	23%	14%

Table 5.2 TDN, DOC and UV analysis of NOM isolates

NOM Isolate	TDN (μgN/L)	RSD (%)	DOC (mgC/L)	UV_{254} (cm^{-1})	SUVA ($[mg/L]^{-1}m^{-1}$)	C/N measured	C/N elemental analysis
Loire Summer HPI	1109	4.74	6.4	0.103	1.61	5.77	7.34
Suwannee TPIA	201	9.68	7.9	0.289	3.66	39.19	46.2
Suwannee TPIN	332	13.09	8.81	0.267	3.03	26.51	29.7
Loire HPO	495	3.89	13.47	0.456	3.39	27.2	24.5
Loire summer HPO	508	1.43	6.37	0.201	3.15	12.54	19.1
Britanny HPO	736	10.84	17.54	0.664	3.79	23.82	-
Gartempe HPO	311	3.16	8.2	0.441	5.38	26.4	24.84
Saint Julien HPO	498	1.79	11.03	0.270	2.45	22.14	17.8
Ribou HPO	443	6.56	12.27	0.242	1.97	27.67	24.9
Seine HPO	473	2.64	9.62	0.255	2.65	20.32	19.8
SPR HPOA	213	3.54	9.56	0.278	2.91	44.86	44.11
Loire summer TPIA	841	15.95	9.16	0.191	2.08	10.9	9.04
Gartempe TPI	587	12.01	10.89	0.395	3.63	18.55	16.6
Loire TPI	1010	6.12	10.39	0.243	2.34	10.28	12
Britanny TPI	847	8.39	9.69	0.173	1.79	11.44	-
Ribou TPI	1139	1.4	8.55	0.190	2.22	7.51	10.2
Seine TPI	838	2.03	11.59	0.149	1.29	13.83	10.6
SPR TPIN	1787	1.9	7.41	0.075	1.01	4.15	4
NF Biofilm	653	0.63	6.53	0.073	1.12	9.99	10.01
Cazeau Lake Colloid	399	4.97	7.16	0.058	0.81	17.93	-
Naintré TPI	1065	1.89	10.84	0.200	1.85	10.18	9.3
Naintré HPO	728	3.49	8.96	0.218	2.43	12.30	13
Ribou Colloid	1003	2.88	8.50	0.088	1.04	8.47	11.4
St Julien Colloid	1138	2.97	5.20	0.113	2.21	4.56	7.9
St Julien TPI	1119	1.1	17.25	0.489	2.83	15.42	-
Britanny Colloid	858	4.2	10.47	0.012	0.11	12.20	-
Cébron Colloid	611	2.26	4.686	0.093	1.98	7.67	-
Naintré Colloid	1315	5.15	11.79	0.245	2.08	5.17	6.9

Table 5.3 Total dissolved amino acids of NOM fractions

	Britanny HPO	Britanny TPI	Gartempe HPO	Gartempe TPI	Loire Summer HPO	Loire Summer HPI	Loire Summer TPI
	μgN/mgC	μgN/mgC	μgN/mgC	μgN/mgC	μgN/mgC	μgN/mgC	μgN/mgC
Galactosamine	-	0.10±0.01	0.03±0.00	0.12±0.03	0.01±0.00	0.04±0.01	0.17±0.04
Glucosamine	-	0.13±0.01	0.10±0.03	0.15±0.03	0.01±0.00	0.02±0.00	0.22±0.05
Aspartic acid	0.34±0.05	0.80±0.07	0.15±0.03	0.58±0.10	0.34±0.11	0.77±0.15	0.96±0.16
Serine	0.13±0.02	0.32±0.11	0.08±0.01	0.26±0.02	0.09±0.02	0.14±0.04	0.44±0.14
Glutamic acid	0.32±0.08	0.76±0.10	0.08±0.01	0.45±0.06	0.26±0.05	0.44±0.13	1.06±0.28
Glycine	0.84±0.11	3.60±0.38	0.50±0.13	1.15±0.14	0.60±0.09	1.81±0.37	3.20±1.22
Histidine	0.05±0.01	0.06±0.01	-	0.11±0.01	0.03±0.01	0.09±0.03	0.16±0.04
Arginine	0.04±0.01	0.09±0.02	0.11±0.02	0.03±0.01	0.14±0.00	0.24±0.06	0.12±0.04
Threonine	0.02±0.00	0.07±0.02	0.10±0.02	0.06±0.01	0.08±0.02	0.10±0.02	0.23±0.04
Alanine	0.06±0.01	0.82±0.22	0.30±0.04	0.18±0.07	0.19±0.06	1.55±0.24	1.17±0.18
Proline	0.18±0.04	0.83±0.13	0.39±0.03	0.15±0.02	-	0.18±0.03	0.39±0.05
Cystine	-	-	0.18±0.02	-	-	-	-
Tyrosine	0.02±0.00	0.04±0.01	0.02±0.00	0.01±0.00	0.02±0.00	0.03±0.02	0.02±0.01
Valine	0.02±0.00	0.05±0.01	0.03±0.00	0.03±0.00	0.03±0.01	0.05±0.01	0.06±0.01
Methionine	0.01±0.00	0.02±0.00	0.07±0.01	0.01±0.00	0.01±0.00	0.06±0.02	0.05±0.01
Ornithine	0.01±0.00	0.02±0.00	-	-	-	-	-
Lysine	0.15±0.02	0.23±0.03	0.22±0.03	0.12±0.02	-	0.15±0.01	0.26±0.09
Isoleucine	0.01±0.00	0.02±0.00	-	0.01±0.00	0.01±0.01	0.03±0.01	0.07±0.01
Leucine	0.03±0.00	0.06±0.01	0.08±0.02	0.03±0.00	0.03±0.01	0.07±0.02	0.07±0.03
Phenylalanine	0.01±0.00	0.03±0.00	0.02±0.01	0.01±0.00	0.01±0.00	0.02±0.00	0.03±0.01

(continued)

Table 5.3 (continued)

	NF Biofilm	Ribou HPO	Ribou TPI	SPR HPO-A	SPR TPI-N	Seine HPO	Seine TPI
	µgN/mgC	µgN/mgC	µgN/mgC	µgN/mgC	µgN/mgC	µgN/mgC	µgN/mgC
Galactosamine	0.58±0.04	-	0.05±0.07	-	-	0.05±0.05	0.30±0.03
Glucosamine	3.38±0.10	0.02±0.00	0.16±0.20	0.04±0.00	0.05±0.05	0.07±0.05	0.71±0.04
Aspartic acid	0.75±0.02	0.21±0.05	0.17±0.06	0.17±0.01	2.77±0.58	0.47±0.53	0.83±0.13
Serine	0.26±0.02	0.10±0.04	0.11±0.08	0.09±0.01	0.70±0.48	0.85±0.95	1.19±0.21
Glutamic acid	1.81±0.21	0.19±0.07	0.09±0.07	0.17±0.01	3.46±0.66	0.80±0.93	1.16±0.18
Glycine	0.30±0.04	0.64±0.09	0.84±0.23	0.63±0.02	23.83±1.33	1.81±1.81	3.97±0.49
Histidine	4.34±0.47	-	-	-	0.19±0.08	-	-
Arginine	0.26±0.02	0.27±0.07	0.96±0.59	0.34±0.03	11.46±2.11	0.35±0.44	0.27±0.06
Threonine	0.47±0.02	0.16±0.01	0.12±0.02	0.22±0.01	0.26±0.01	0.30±0.23	0.99±0.24
Alanine	1.49±0.03	0.28±0.05	0.52±0.05	0.30±0.02	4.82±0.30	1.22±1.01	2.76±0.41
Proline	0.15±0.01	0.25±0.06	0.34±0.06	0.02±0.01	1.49±0.18	0.56±0.59	0.91±0.03
Cystine	-	0.05±0.01	0.09±0.02	0.04±0.01	1.55±0.08	0.03±0.01	0.07±0.04
Tyrosine	-	0.06±0.01	0.04±0.01	0.05±0.01	0.07±0.02	0.30±0.30	0.25±0.05
Valine	0.14±0.01	0.16±0.04	0.05±0.01	-	0.07±0.06	0.85±0.80	1.17±0.13
Methionine	0.09±0.01	0.01±0.01	0.05±0.02	-	0.24±0.21	0.19±0.05	0.22±0.01
Ornithine	na	na	Na	na	na	na	na
Lysine	0.64±0.43	0.30±0.04	0.22±0.02	-	0.36±0.04	0.81±0.78	0.64±0.17
Isoleucine	-	0.03±0.07	0.01±0.11	0.03±0.05	0.28±2.73	0.81±0.83	0.82±0.14
Leucine	-	0.08±0.02	0.07±0.01	0.08±0.00	0.81±0.82	1.65±1.84	1.54±0.32
Phenylalanine	0.10±0.01	0.07±0.01	0.02±0.00	0.06±0.00	0.91±0.09	0.87±0.96	0.95±0.20

(continued)

Table 5.3 (continued)

	Saint Julien HPO µgN/mgC	Suwannee TPIA µgN/mgC	Suwannee TPIN µgN/mgC	Loire HPO µgN/mgC	Loire TPI µgN/mgC
Galactosamine	0.02±0.02	0.05±0.02	0.07±0.02	-	0.04±0.01
Glucosamine	0.03±0.01	0.07±0.02	0.07±0.02	-	-
Aspartic acid	0.16±0.03	0.28±0.04	0.65±0.14	0.14±0.03	0.51±0.04
Serine	0.25±0.07	0.21±0.05	0.31±0.07	0.04±0.01	0.09±0.01
Glutamic acid	0.30±0.06	0.22±0.01	0.72±0.17	0.13±0.04	0.48±0.06
Glycine	0.76±0.15	0.76±0.11	3.50±0.97	0.34±0.06	7.60±0.66
Histidine	-	0.06±0.04	0.12±0.04	-	0.03±0.00
Arginine	0.08±0.03	0.02±0.01	0.08±0.02	0.07±0.03	0.56±0.12
Threonine	0.15±0.01	0.02±0.01	0.03±0.01	0.03±0.01	0.19±0.03
Alanine	0.43±0.05	0.04±0.01	0.26±0.09	0.15±0.03	1.57±0.09
Proline	0.18±0.01	0.22±0.04	0.77±0.20	-	0.82±0.09
Cystine	0.01±0.01	-	-	-	-
Tyrosine	1.23±0.06	0.04±0.01	0.02±0.00	-	0.06±0.02
Valine	0.35±0.03	0.01±0.00	0.03±0.01	0.02±0.01	0.12±0.01
Methionine	0.18±0.01	-	-	-	0.05±0.01
Ornithine	Na	-	-	-	-
Lysine	0.31±0.05	0.10±0.04	0.04±0.01	-	-
Isoleucine	0.34±0.03	0.01±0.00	0.01±0.00	0.01±0.00	0.02±0.00
Leucine	0.62±0.09	0.01±0.00	0.03±0.01	0.03±0.01	0.09±0.02
Phenylalanine	0.39±0.05	-	0.02±0.00	0.01±0.00	0.03±0.00

(continued)

Table 5.3 (continued)

	Ribou Colloid	Britanny Colloid	Saint Julien TPI	Saint Julien Colloid	Naintre HPO	Naintre TPI	Naintre colloid	Cazeau Lake colloid	Cebron colloid
	µgN/mgC	µgN/mgC	µgN/mgC	µgN/mgC	µgN/mgC	µgN/mgC	µgN/mgC	µgN/mgC	µgN/mgC
Galactosamine	0.30±0.10	2.75±0.18	0.18±0.05	1.67±0.38	-	-	3.06±0.29	0.80±0.14	0.77±0.22
Glucosamine	0.33±0.12	5.09±0.60	0.18±0.05	3.24±0.26	-	0.22±0.03	4.79±0.39	0.82±0.22	0.53±0.16
Aspartic acid	0.57±0.04	1.55±0.23	0.93±0.08	6.67±0.76	0.63±0.15	0.81±0.09	5.99±0.52	1.65±0.59	1.73±0.16
Serine	0.67±0.04	2.59±0.30	0.51±0.08	4.05±0.59	0.48±0.15	0.47±0.06	4.05±0.26	1.92±0.12	1.58±0.20
Glutamic acid	0.45±0.06	1.23±0.30	0.90±0.09	5.91±0.80	1.01±0.25	1.05±0.14	4.88±0.50	1.40±0.37	1.46±0.19
Glycine	0.97±0.10	2.88±0.39	2.37±0.22	8.03±1.11	1.62±0.40	2.12±0.23	6.85±0.57	2.63±0.59	3.03±0.32
Histidine	0.17±0.02	0.38±0.11	-	2.12±0.39	0.39±0.08	0.49±0.10	1.50±0.24	-	0.35±0.08
Arginine	-	0.62±0.12	0.43±0.04	6.88±0.69	0.92±0.24	0.41±0.10	6.32±0.79	1.13±0.44	-
Threonine	0.37±0.03	2.73±0.14	0.64±0.19	5.02±0.27	0.25±0.06	0.34±0.03	5.02±0.59	1.99±0.41	1.16±0.08
Alanine	0.62±0.04	3.51±0.24	0.83±0.06	7.96±0.48	5.61±1.18	0.82±0.14	8.33±0.96	3.01±0.74	2.43±0.32
Proline	0.89±0.11	-	-	1.92±0.28	0.17±0.03	0.47±0.08	1.82±0.23	0.59±0.10	2.99±0.47
Cystine	-	-	-	-	-	-	-	-	-
Tyrosine	-	0.39±0.05	0.07±0.00	1.58±0.22	0.22±0.02	0.12±0.03	1.47±0.12	0.26±0.11	0.27±0.06
Valine	0.09±0.01	1.14±0.15	0.24±0.04	4.14±0.21	0.24±0.06	0.31±0.07	3.92±0.33	1.03±0.33	0.50±0.05
Methionine	0.07±0.01	0.12±0.02	-	0.72±0.07	0.08±0.02	0.05±0.01	0.72±0.09	0.11±0.02	-
Lysine	0.30±0.01	0.89±0.14	0.34±0.19	4.86±0.49	0.49±0.18	0.50±0.11	4.52±0.69	1.09±0.32	0.85±0.15
Isoleucine	0.04±0.01	0.57±0.10	0.10±0.01	2.29±0.20	0.18±0.05	0.16±0.01	2.22±0.17	0.40±0.11	0.15±0.03
Leucine	0.08±0.01	1.04±0.24	0.21±0.03	4.49±0.28	0.31±0.07	0.25±0.04	4.03±0.42	1.08±0.39	0.37±0.06
Phenylalanine	0.05±0.00	0.54±0.13	0.09±0.04	1.98±0.15	0.17±0.05	0.10±0.02	1.77±0.20	0.55±0.16	0.17±0.03

Table 5.4 Amino acids analysis of the Saint Julien Colloid Fraction after MSA or HCl hydrolysis using either AccQTag or OPA detection method.

	HCl Hydrolysis AccQTag (μM)	MSA hydrolysis OPA (μM)
Asp	2.44±0.28	4.43±2.55
Glu	2.16±0.29	2.91±0.77
Ser	1.48±0.21	1.32±0.28
His	0.26±0.05	0.35±0.10
Arg	0.63±0.06	4.18±0.71
Gly	2.94±0.40	1.48±0.35
Ala	2.91±0.18	4.83±1.09
Tyr	0.58±0.08	0.81±0.16
Met	0.26±0.03	0.24±0.09
Val	1.52±0.08	1.59±0.76
Phe	0.73±0.06	0.61±0.15
Ile	0.84±0.07	1.09±0.28
Leu	1.64±0.10	3.50±0.86
Lys	0.89±0.09	0.59±0.44
Sum AA (μgN/L)	316	584
% Total N*	30	53

*Percentage total N recovered was determined by adding amino sugars measured by the AccQTag method.

Table 5.5 Amino acids analysis of the Saint Julien TPI Fraction after MSA or HCl hydrolysis and using AccQTag or OPA detection method.

	HCl hydrolysis AccQTag (μM)	MSA hydrolysis OPA (μM)
Asp	0.72±0.07	0.51±0.06
Glu	0.70±0.07	0.98±0.09
Ser	0.40±0.06	0.44±0.06
His	-	0.08±0.01
Arg	0.08±0.01	2.75±0.34
Gly	1.84±0.17	0.35±0.03
Ala	0.64±0.04	1.04±0.24
Tyr	0.06±0.00	0.12±0.02
Met	-	0.05±0.01
Val	0.19±0.03	0.37±0.04
Phe	0.07±0.03	0.11±0.01
Ile	0.08±0.01	0.14±0.02
Leu	0.16±0.03	0.42±0.01
Lys	0.13±0.07	-
Sum AA (μgN/L)	75	221
%total N*	8	21

*Percentage total N recovered was determined by adding amino sugars measured by the AccQTag method. Amino sugars only accounted for 0.35% of the total N.

Table 5.6 Bulk sample data from the South Platte River study segment

Site Name	DOC (mg/L)		SUVA (L/mg-m)		DON (mgN/L)		DOC/DON (mgC/mgN)	
	April	May	April	May	April	May	April	May
Chat R	4.4	4.1	1.5	1.9	0.26	0.27	17	15
SP@285	6.1	5.2	2.0	1.8	0.11	0.22	55	24
L/E Eff	9.2	10.1	1.5	1.4	BDL	BDL	--	--
SP @ Burl	6.0	5.2	1.9	2.1	0.12	0.07	50	74
M-2 Eff	13.9	12.4	1.2	1.1	2.65	1.58	5	8
M-N/D Eff	10.9	10.5	1.6	1.5	1.19	0.26	9	40
SP@ SC	10.7	9.4	1.4	1.4	0.14	0.92	76	10
SP @ Thor	8.7	9.3	1.7	1.4	0.14	0.21	62	44

Table 5.7 URI values from SEC-DOC chromatograms

Site Name	High Flow URI Values				Low Flow URI Values			
	Peak 1 (2000)	Peak 2 (3000)	Peak 3 (3500)	Peak 4 (4000)	Peak 1 (2000)	Peak 2 (3000)	Peak 3 (3500)	Peak 4 (4000)
Chat R	4.0	1.9	1.9		6.0	2.0	2.0	4.0
SP @ 285		2.0		3.9	6.0	2.0	2.0	4.8
L/Eeff	7.0	2.3	1.8		10.6	2.6	2.0	
SP @ Burl	6.5		2.0	7.3	6.0	2.0		10.2
M-2Eff	9.0	3.1	2.5	5.2	8.5	3.0	2.6	5.6
M-N/Deff	7.0	2.5	2.0		11.8	2.7	2.5	11.9
SP @ SC	7.0	2.5	2.0	9.0	9.8	2.6	2.4	7.4
SP @ Thor	13.0	2.4	2.1	10.4	8.6	2.6	2.4	10.6

Table 5.8 Analysis of samples from the South Platte River collected during high flow

	Organic Carbon		Pyrolysis GC/MS fractions (% in each grouping)				
Sample	Initial DOC (mg/L)	DOC lost during pretreatment	Proteins & Peptides	Aminosugars	Poly-saccharides	Polyhydroxy Aromatic	Miscellaneous
SP @ 285	4.4	77%	3.2	0.0	2.3	1.6	73.4
L/Eeff	7.2	70%	2.3	1.6	8.4	3.1	69.2
SP @ Burl	4.8	54%	2.7	1.3	6.8	4.0	67.5
M-2eff	11.2	65%	2.8	0.6	12.2	7.2	59.6
M-N/Deff	8.7	56%	2.6	0.9	7.7	6.2	65.4
SP@SC	8.7	57%	1.6	1.0	6.8	3.5	75.9
SP @ Thor	6.3	35%	2.0	0.6	11.3	4.1	68.9

Table 5.9 BSA concentrations in solutions with different ion types and concentrations

Ion	Concentration (mM)	Cation (mg/L)	Anion (mg/L)	mg/L BSA	
				Mean	RSD
NaCl	1	23	35.5	2.96	0.16
	5	115	177.5	3.03	0.14
	10	230	355	3.49	0.05
Na_2SO_4	1	46	96	3.04	0.02
	5	230	480	3.41	0.03
	10	460	960	3.83	0.16
$CaCl_2$	0.5	20	35.5	5.18	0.44
	1	40	71	4.01	0.08
	2	80	142	4.12	0.18
$NaHCO_3$	1	23	61	2.98	0.03
	5	115	305	3.08	0.16
	10	230	610	3.21	0.02

Table 5.10 Percent protein in DON for several NOM isolates

Source water grouping for NOM isolates	Protein / DON (%)	
	Mean	RSD
Lake/Reservoir (n = 9)*	15	5
River (n = 10)*	5	3
Finished drinking water (n = 2)*	3	1
Colloids (n = 4)	42	7

* Includes HPO, TPI, and HPI fractions; colloid fraction was excluded.

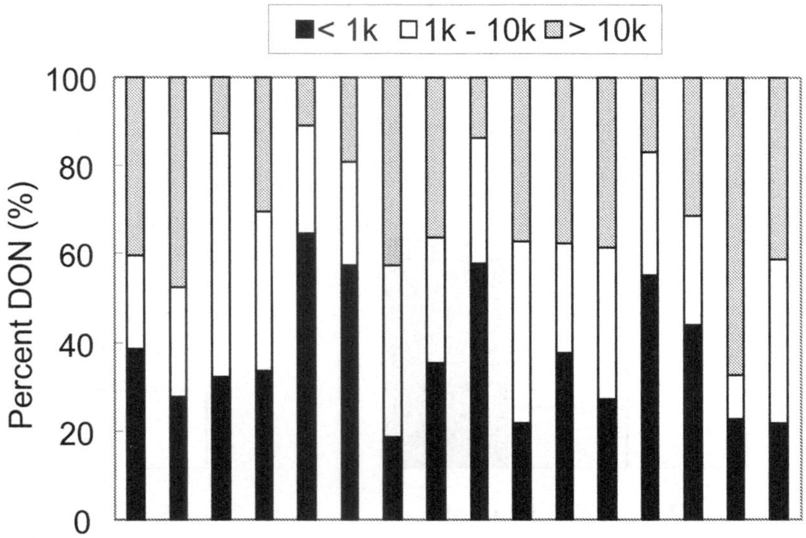

Figure 5.1 Apparent molecular weight distribution based upon ultrafiltration for DON in 16 raw drinking water samples

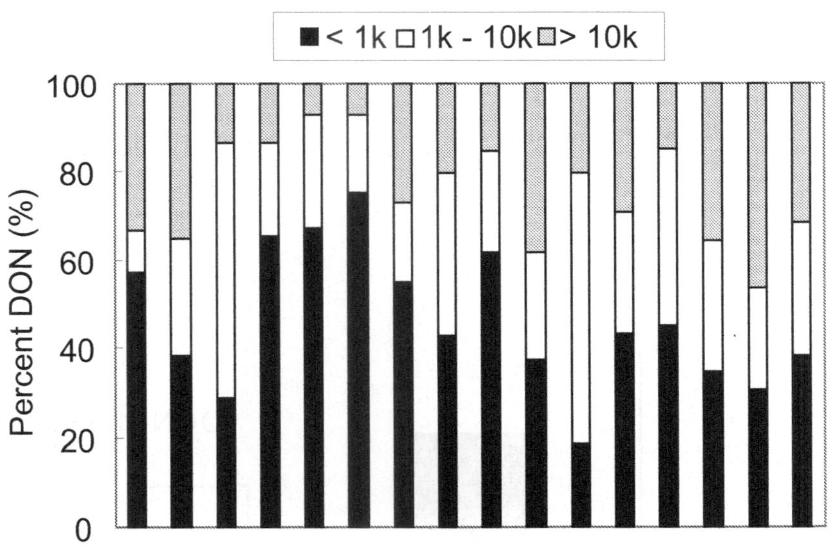

Figure 5.2 Apparent molecular weight distribution based upon ultrafiltration for DON in 16 finished drinking water samples

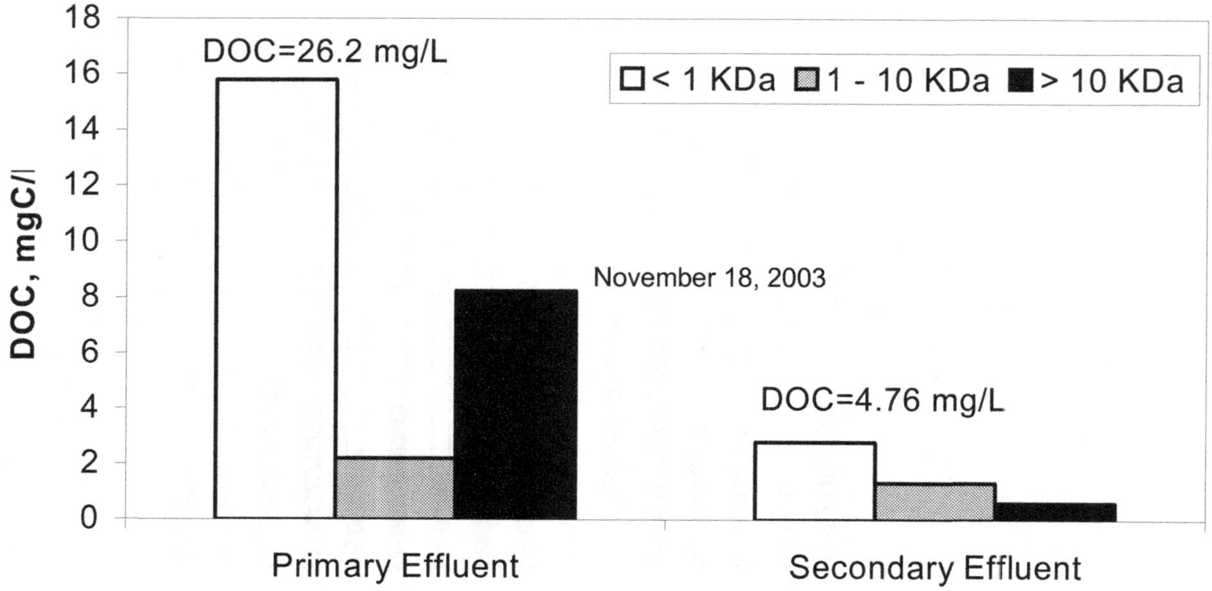

Figure 5.3 MW distributions of DOC in primary effluent and secondary effluent from the Mesa, Arizona WWTP

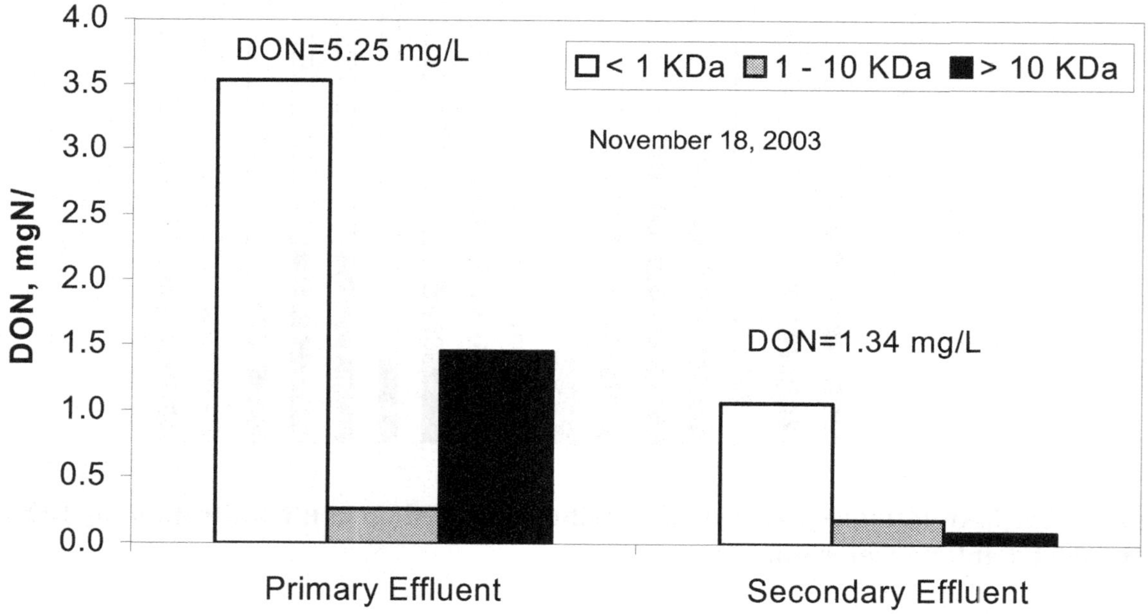

Figure 5.4 MW distributions of DON in primary effluent and secondary effluent from the Mesa, Arizona WWTP

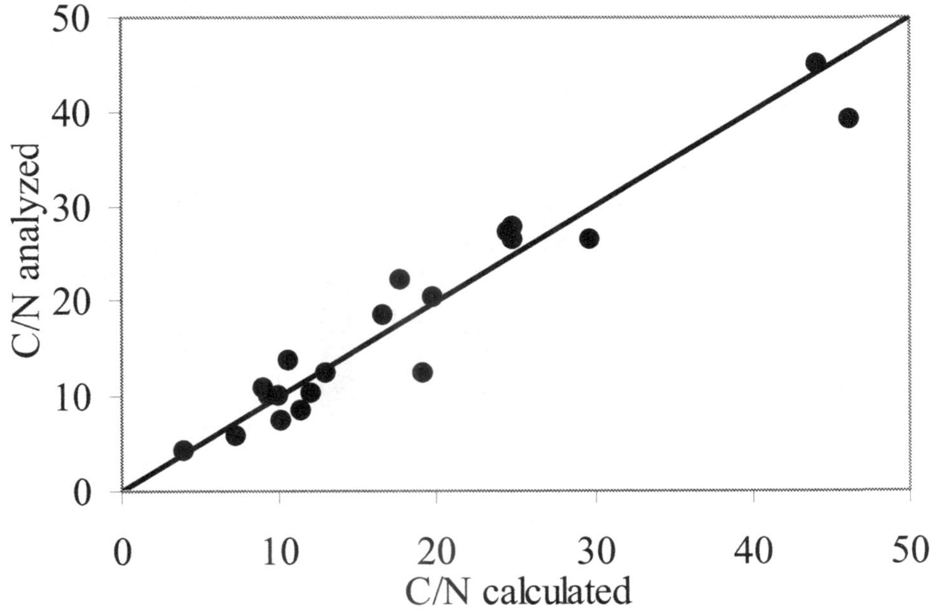

Figure 5.5 Ratio of carbon to nitrogen based upon elemental analysis (C/N calculated) and high temperature combustion (C/N analyzed) of samples

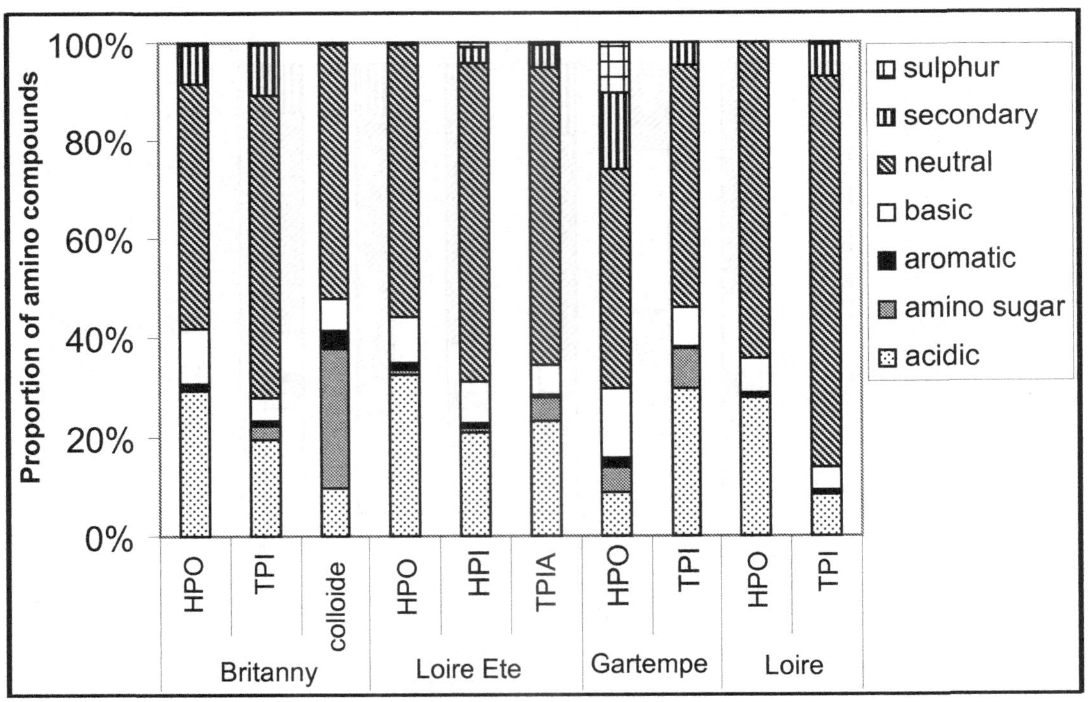

Figure 5.6 Relative distribution of the different classes of amino acids and aminosugars for NOM isolates

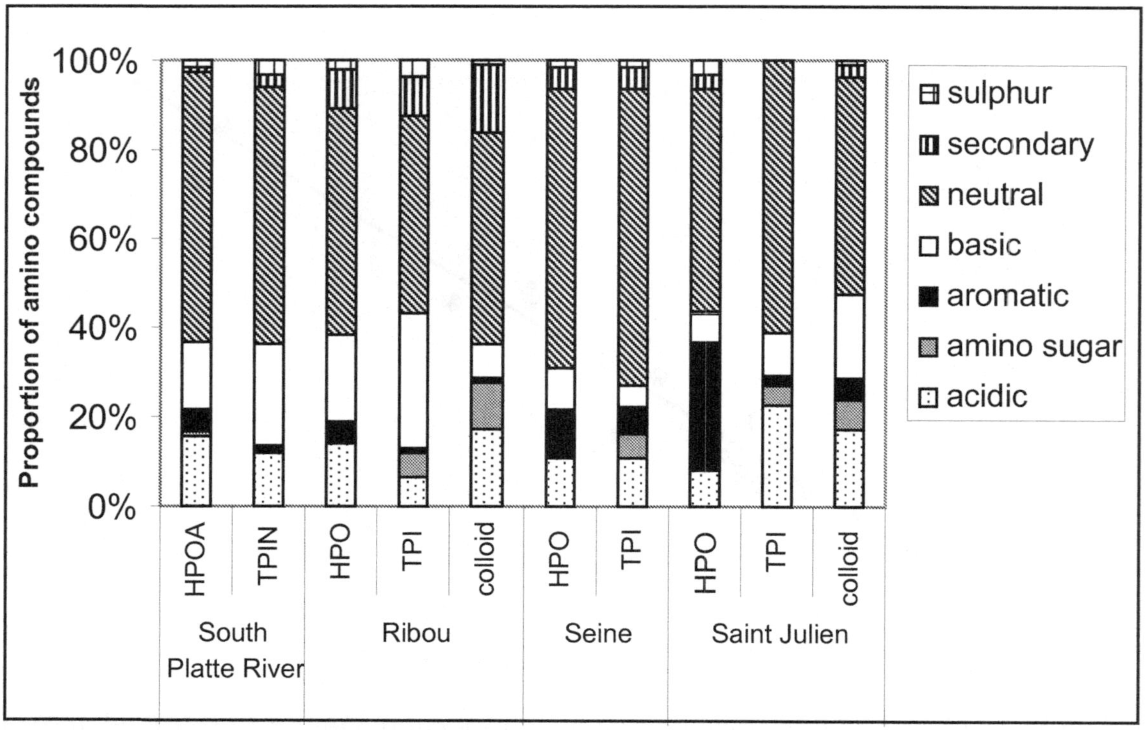

Figure 5.7 Relative distribution of the different classes of amino acids and aminosugars for NOM isolates

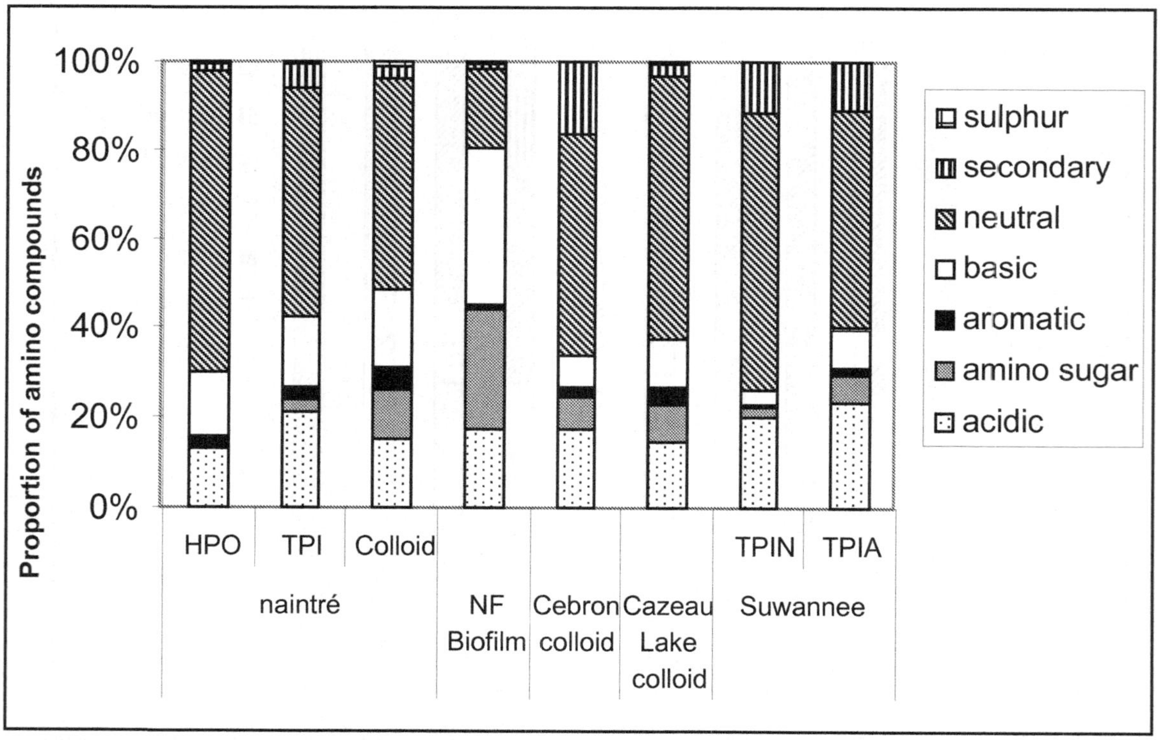

Figure 5.8 Relative distribution of the different classes of amino acids and aminosugars for NOM isolates

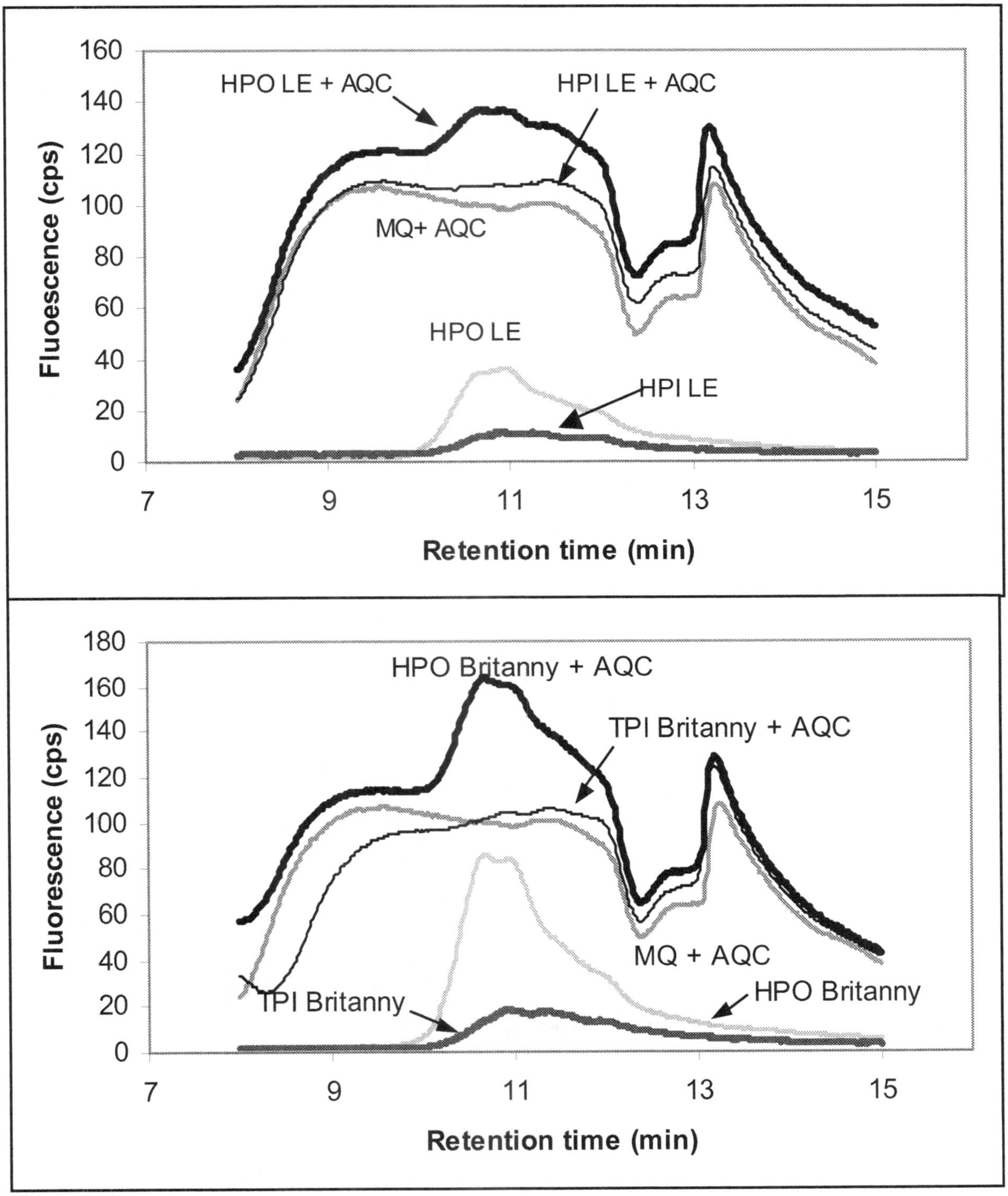

Figure 5.9 HPSEC/fluorescence chromatograms of NOM isolate solutions after AQC derivatization

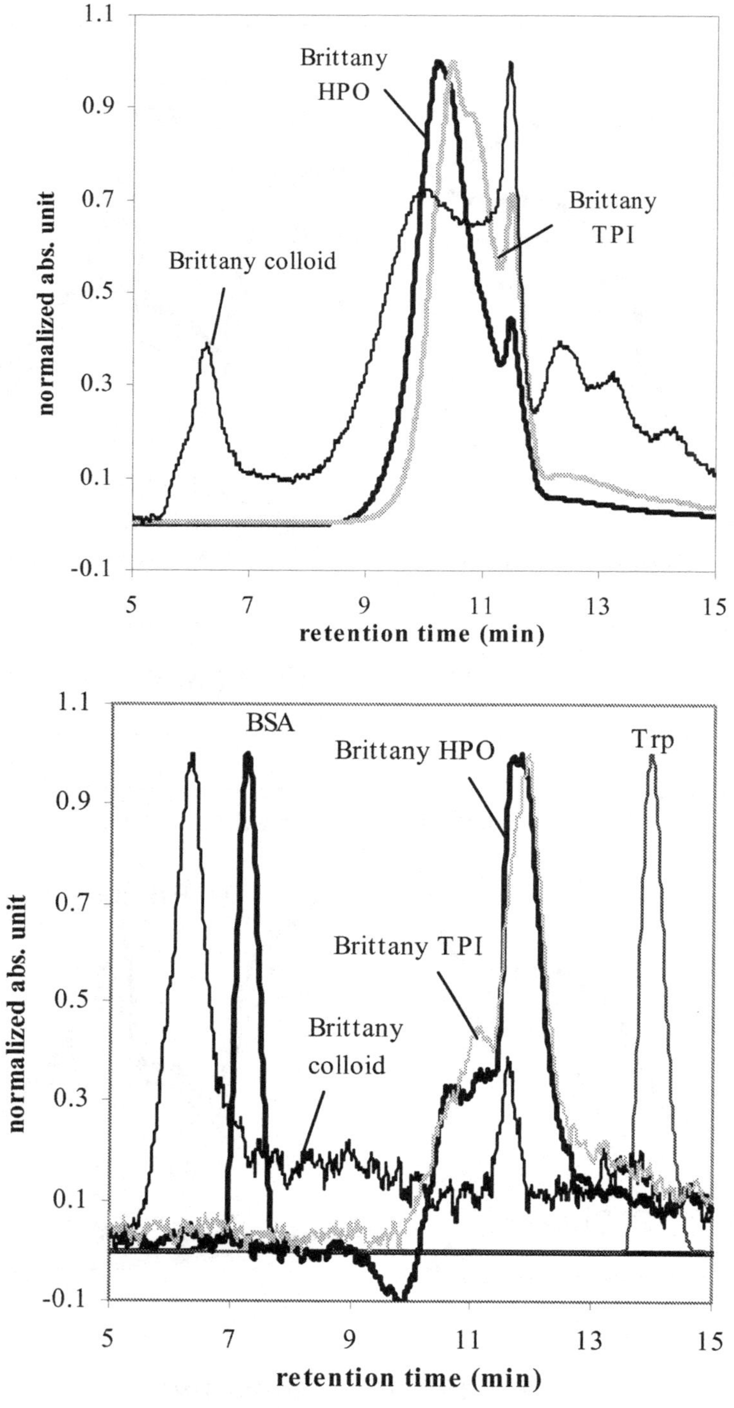

Conditions: HPSEC/fluorescence profiles of BSA protein and tryptophane amino acid are also given for comparison. Chromatograms on left are with UV detection at 260 nm and on the right are chromatograms with fluorescence detection (λ_{exc}: 280 nm, λ_{em}: 320 nm).

Figure 5.10 HPSEC chromatograms of the colloid, HPO and TPI fractions isolated from a river in Britany, France.

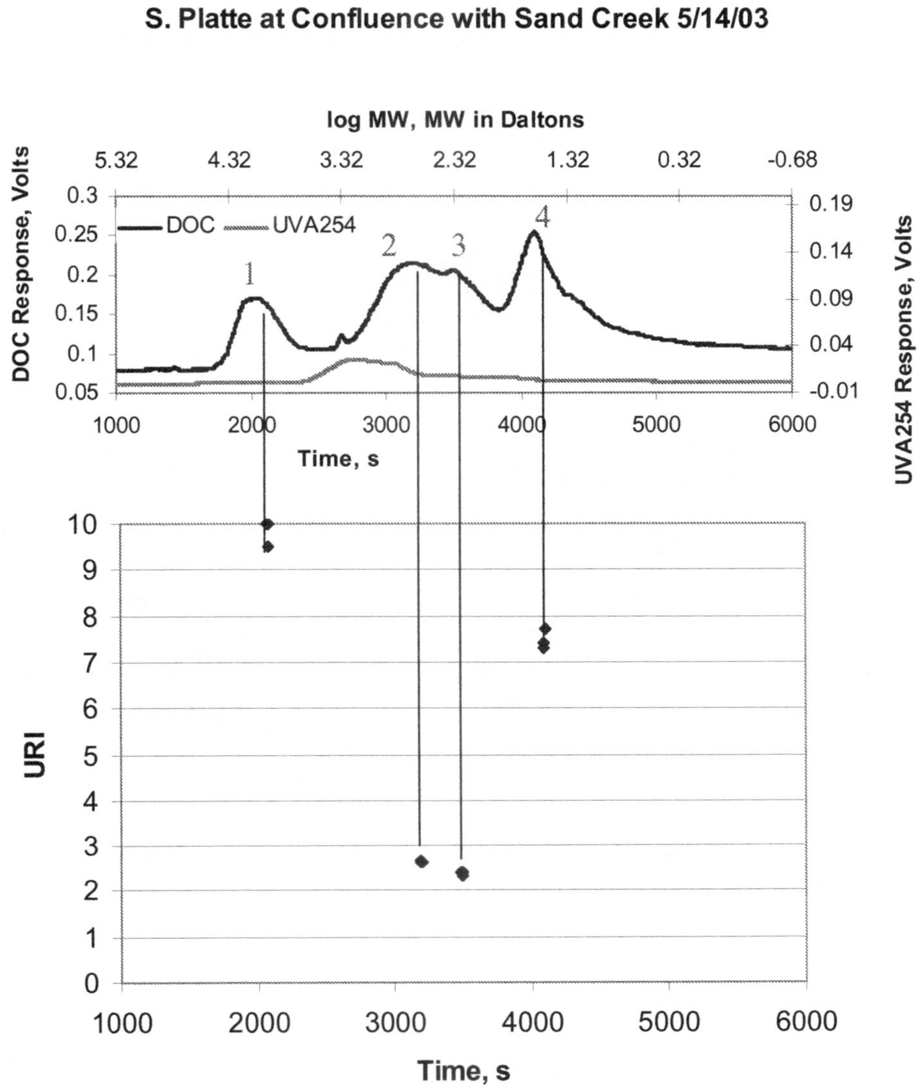

Figure 5.11 Determination of URI values next to an HPSEC with DOC detection

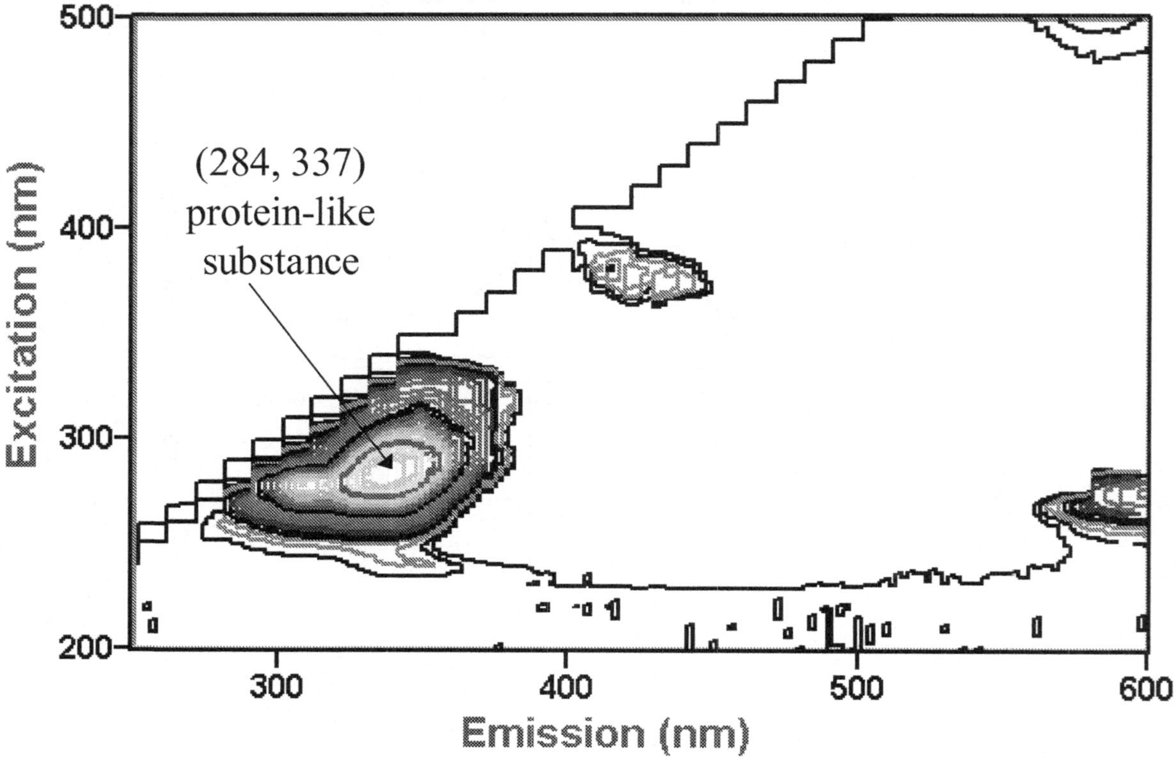

Figure 5.12 Differential EEM fluorescence spectra for a sampling site on the South Platte River [high flow EEM subtracted from low flow EEM].

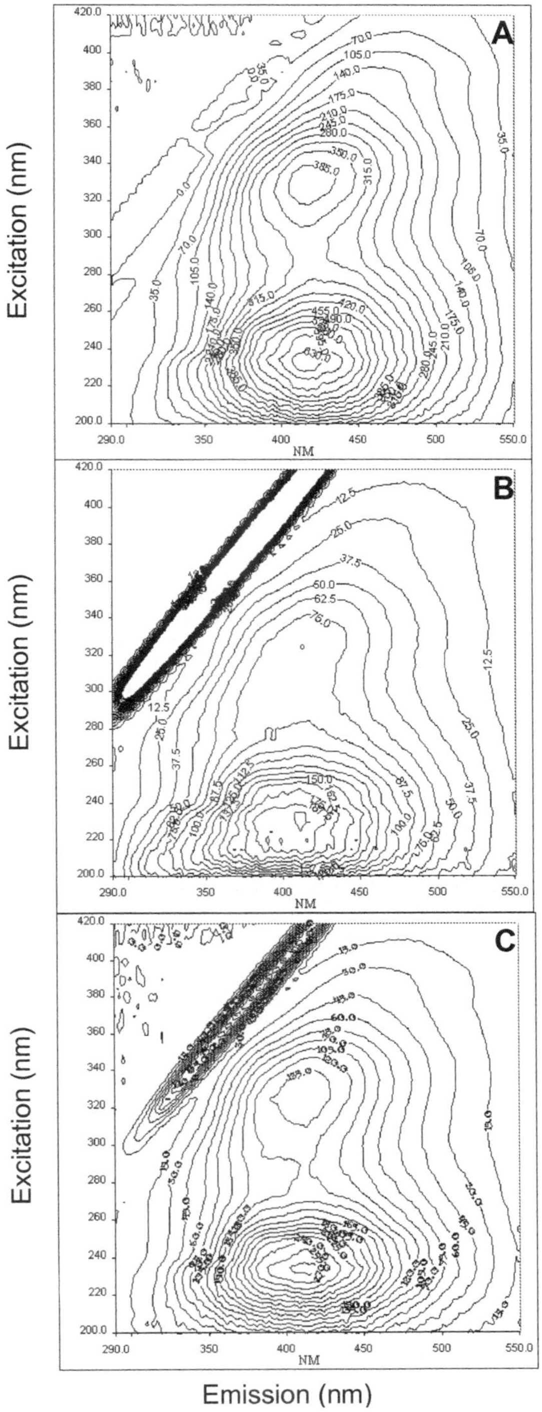

Figure 5.13 EEMs for raw water from A) Ann Arbor, MI [DON=0.295 mgN/L], B) Deer Valley, AZ [DON = 0.166 mgN/L, and C) Southborough, MA [DON=0.079 mgN/L]

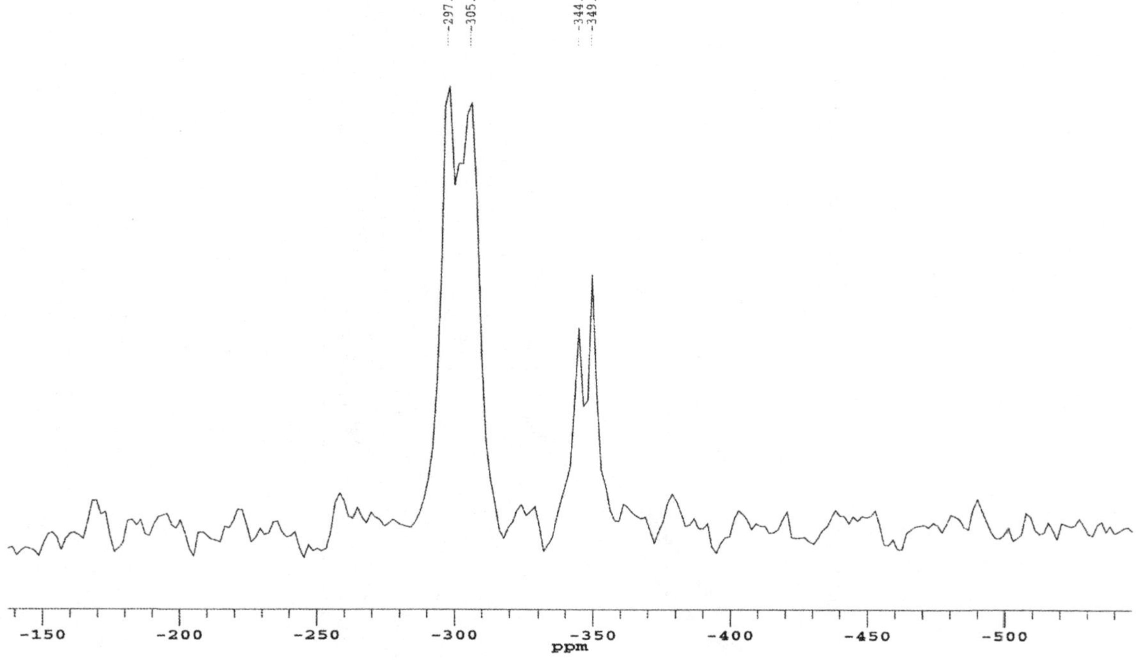

Figure 5.14 ^{15}N-NMR spectra for the IHSS Elliot soil humic acid

CHAPTER 6 DISSOLVED ORGANIC NITROGEN REMOVAL IN BENCH-SCALE TESTS

This chapter evaluates DON removal using bench-scale experiments. Alum and alum plus cationic polymer jar tests were conducted. Limited work on DON adsorption on powder activated carbon (PAC) was also evaluated.

COAGULATION WITH METAL SALTS

Three waters from Arizona were used in jar tests. Central Arizona Project (CAP) water is imported from the Colorado River and Verde River is from a local watershed. The third sample was a collected during a large runoff event on the Salt River, following a severe forest fire, and represented a high-SUVA water. Table 6.1 summarizes the characteristics of these three waters. Alum dosages of up to 100 mg/L were investigated.

Figures 6.1 through 6.3 present jar test results for the three waters, showing DON, DOC, and UVA_{254} removal as a function of alum dose. In the CAP water DON was removed less effectively than DOC, and UVA_{254} had the highest percentage reduction. Coagulation of Verde River or the fire runoff water exhibited comparable DON and DOC removal at high alum dosages. Additional jar tests of secondary treated wastewater resulted in poor removal of DON of effluent organic matter origin.

COAGULATION WITH METAL SALTS AND ORGANIC POLYMERS

Coagulation of several surface waters from around the US was conducted. One representative set of results will be shown, using Huron River, MI water (Table 6.1). Results from other waters followed similar trends. Results are plotted as a function of alum dose for turbidity, DON, DOC, and UVA_{254} in Figures 6.4 through 6.7. Alum alone or in conjunction with cationic polymers achieve excellent turbidity removal. Cationic polymer addition slightly improved DOC and UVA_{254} removals at low alum dosages. Addition of cationic polymer (polyDADMAC contains 7% organic nitrogen by mass) actually increased DON removal at low (reasonable) alum dosages. It is hypothesized that since the more polar NOM fractions are nitrogen-enriched (see Chapter 5), that cationic polymers aid in neutralizing charge associated with organic molecules and/or bridges these polar molecules onto the surfaces of aluminum hydroxides that precipitate. At alum dosages up to 5 mgAlum per mg DOC, cationic polymer improved DON removal by an additional 15% to 20% over alum alone. At very high alum dosages, the effect of cationic polymer addition on DON removal was negligible. Molecular weight fractionation before and after coagulation indicate that cationic polymer addition improved removal of all DON weight fractions. In all cases, it appears that the added DON as cationic polymer was nearly completely removed.

ADSORPTION ON ACTIVATED CARBON

Adsorption of DON onto activated carbon (20 hour contact time) was evaluated on several water matrices. Results from two waters will be discussed (Table 6.2). These waters were selected since the DON concentrations were high and large percentage removals could easily be quantified. Two activated carbon dosages were applied (5 and 25 mg/L). Figures 6.8

and 6.9 presents the results for the CAP water spiked with a NOM isolate and a WWTP effluent sample from Mesa Arizona, respectively. Higher activated carbon dosages increased DON removal in all cases. There was negligible differences in DON removal among activated carbon brands. The DON in the WWTP effluent (50% DON removal) was removed better than DON in the CAP plus XAD-Isolate water, where only 20% of the DON was removed.

SUMMARY

Alum coagulation removed DON, on a percentage of initial concentration basis, at levels comparable with the DOC removal. Coagulation preferentially removed higher molecular weight (>10,000 dalton) DOC and DON materials. Adding cationic polymers improved DON removal to a larger extent than DOC removal. It is hypothesized that organic nitrogen is associated with polar NOM fractions, and that cationic polymer improves neutralization of this polar NOM and aids in its removal during coagulation. NOM of wastewater origin was more difficult to remove during coagulation than from surface waters commonly used as drinking water sources. However, DON present in wastewater could be adsorbed onto activated carbon.

Table 6.1 Water quality parameters for samples used in alum coagulation jar tests

Watershed Source	pH	Alkalinity mg/L $CaCO_3$	DOC mg/L	SUVA $m^{-1}[mg/L]^{-1}$	TDN mg-N/L	DON mg-N/L	DON/TDN Ratio (x100%)
CAP Water	8.4	~ 105	3.4	1.2	0.29	0.25	86%
Verde River	8.4	~ 150	4.3	2.2	0.29	0.28	97%
Fire Runoff	8.0	~ 200	24.8	4.8	2.70	1.71	63%
Huron River	8.2	208	7.4	3.8	0.72	0.33	46%

Table 6.2 Characteristics of water sources

Water source	DON (mgN/L)	DOC (mgC/L)	UVA_{254} (cm^{-1})	SUVA ($Lmg^{-1}m^{-1}$)	DOC/DON
CAP + XAD-isolate	0.70	10.5	0.183	1.7	15.0
WWTP effluent	0.86	5.5	0.092	1.7	6.4

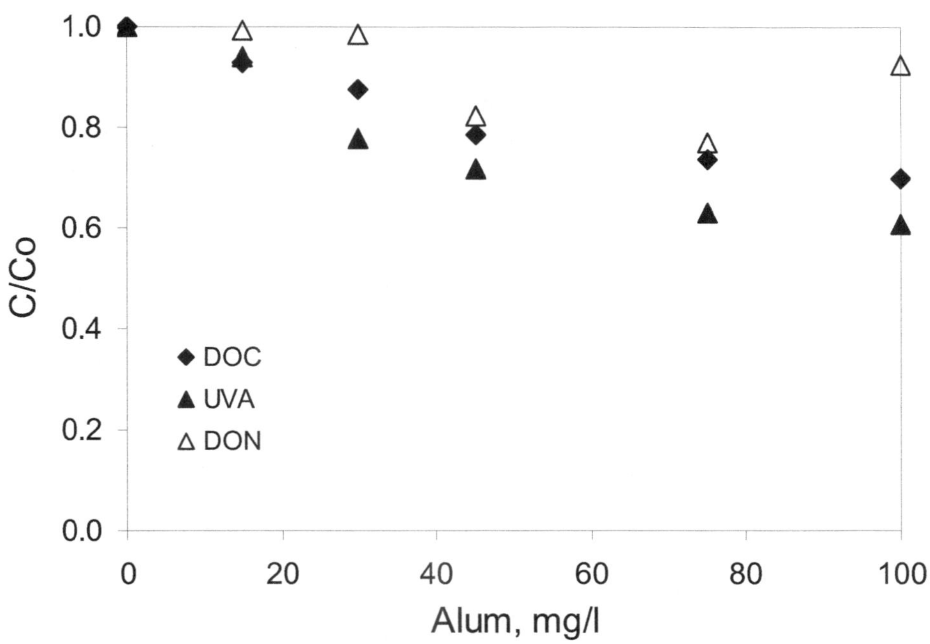

Figure 6.1 Fraction remaining upon coagulation with aluminum sulfate in CAP water sample

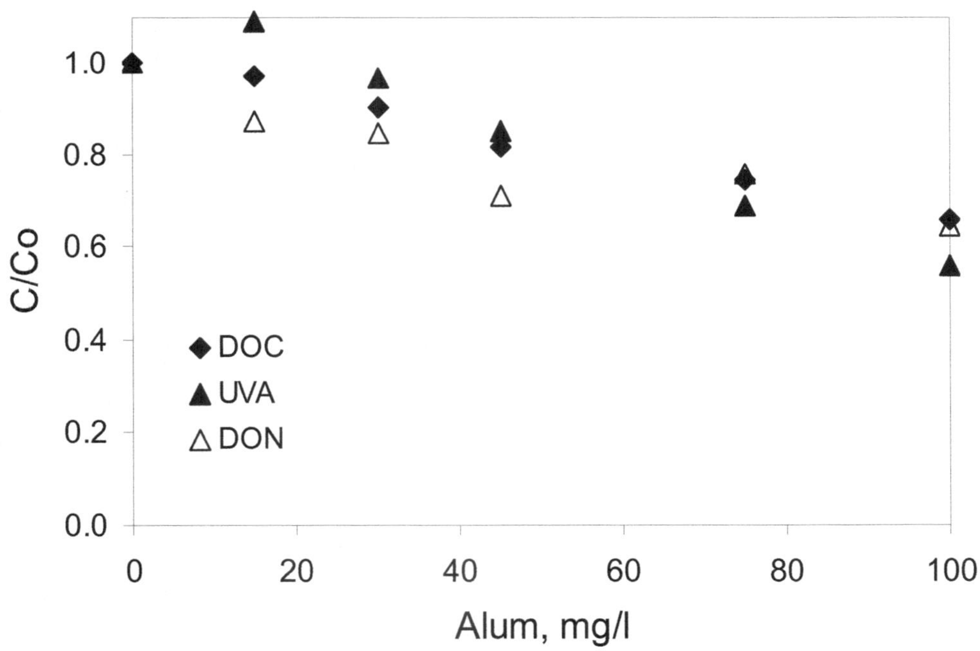

Figure 6.2 Fraction remaining upon coagulation with aluminum sulfate in Verde River water sample

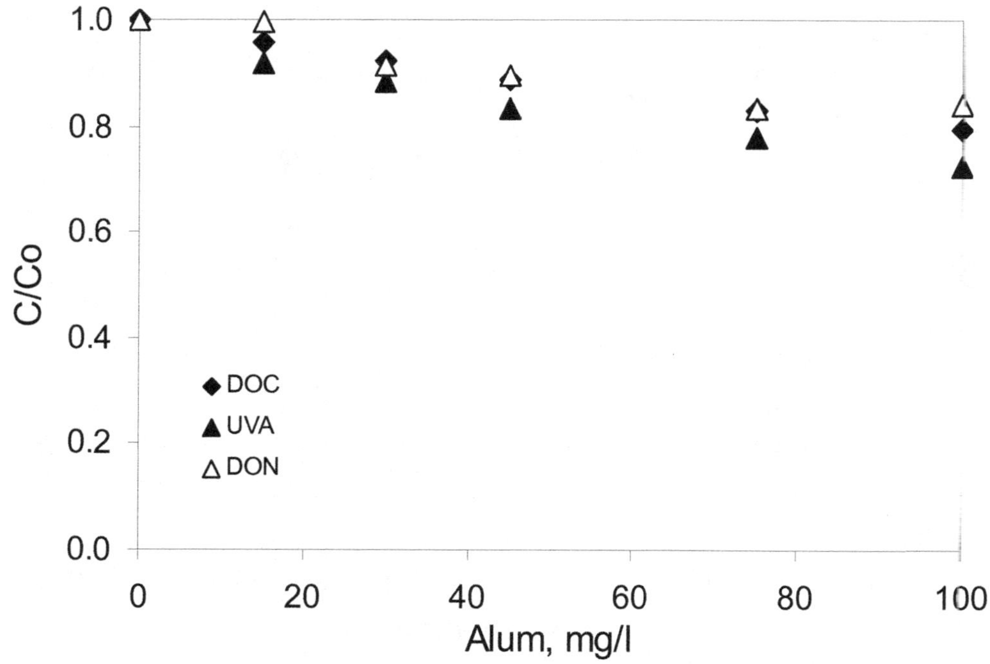

Figure 6.3 Fraction remaining upon coagulation with aluminum sulfate in forest-fire runoff water sample

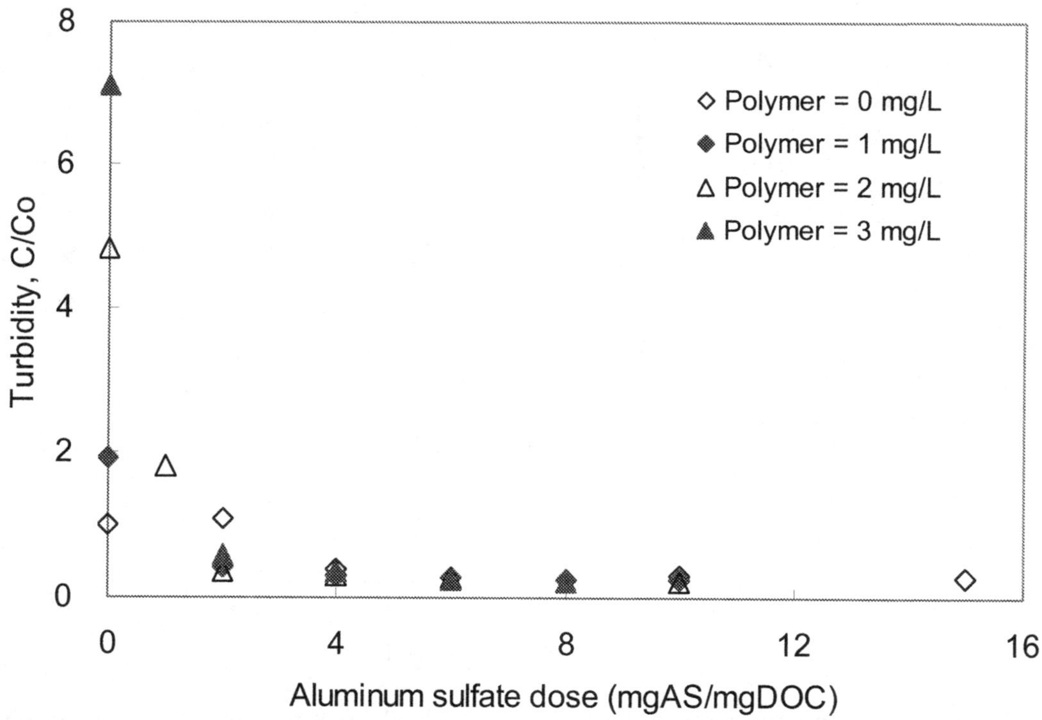

Figure 6.4 Fraction turbidity remaining upon coagulation with aluminum sulfate and cationic polymer with water from the Huron River, MI

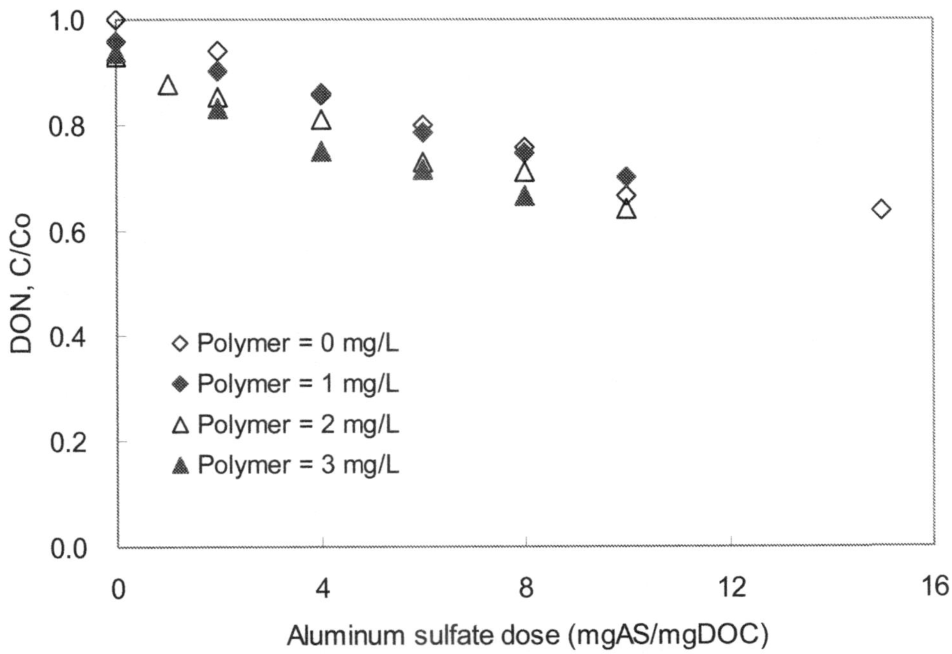

Figure 6.5 Fraction DON remaining upon coagulation with aluminum sulfate and cationic polymer with water from the Huron River, MI

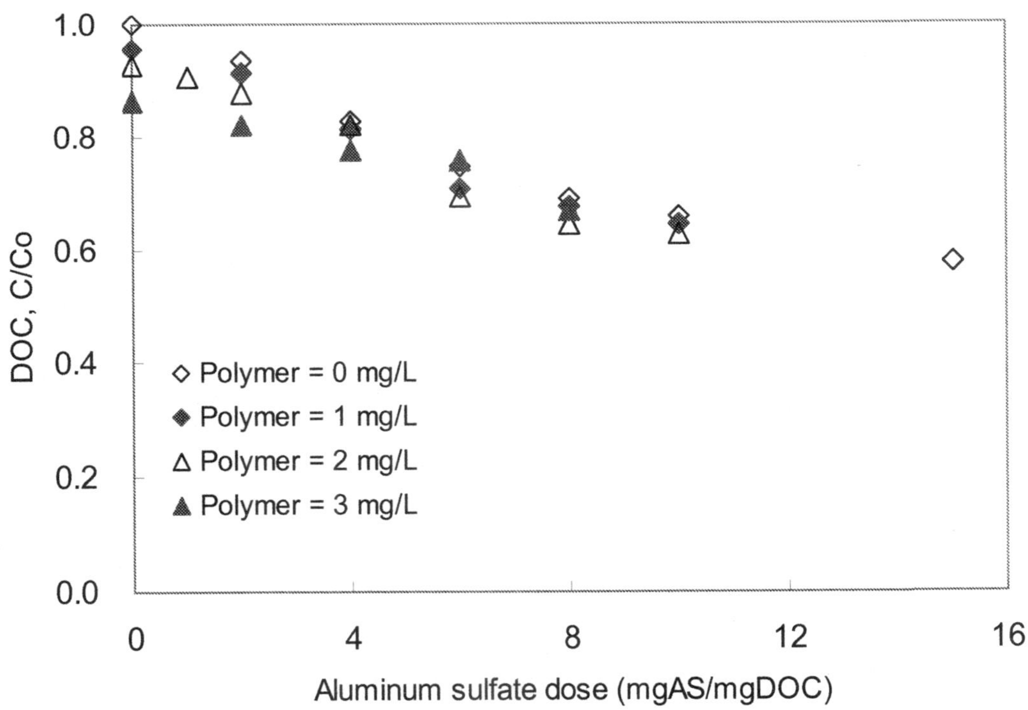

Figure 6.6 Fraction DOC remaining upon coagulation with aluminum sulfate and cationic polymer with water from the Huron River, MI

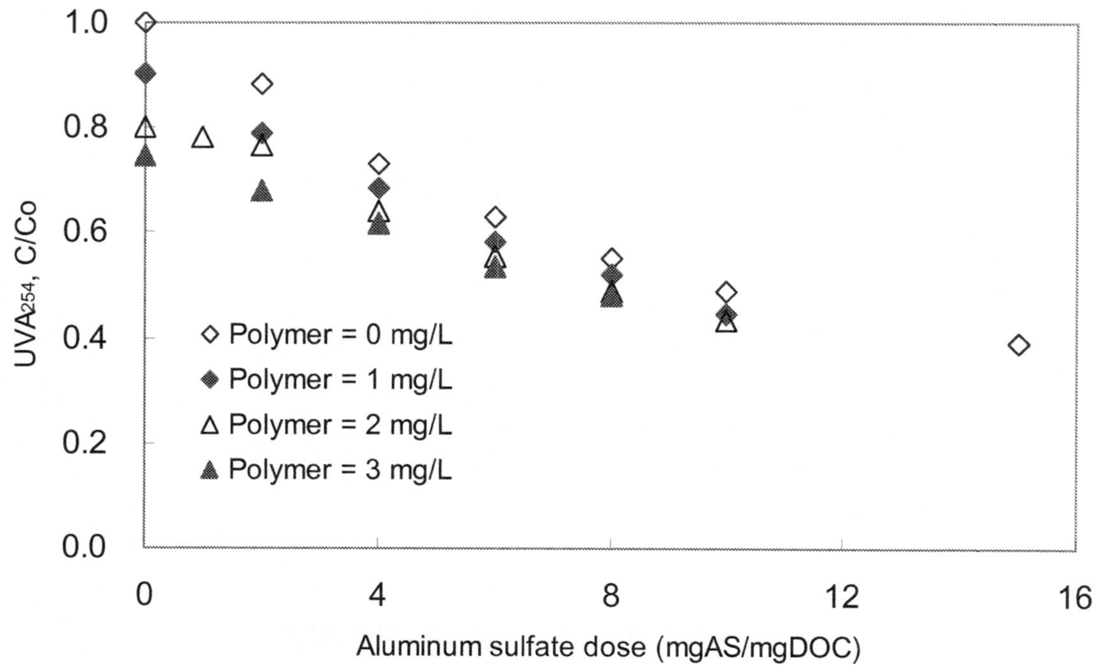

Figure 6.7 Fraction UVA_{254} remaining upon coagulation with aluminum sulfate and cationic polymer with water from the Huron River, MI

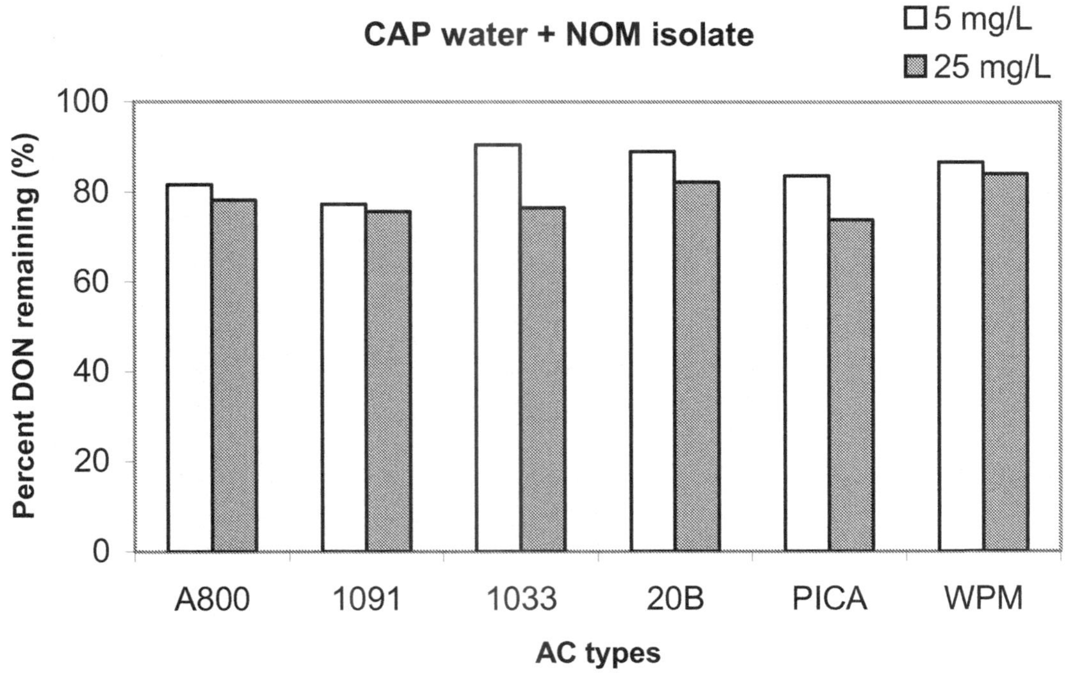

Figure 6.8 DON remaining in CAP water spiked with NOM isolate after PAC adsorption experiments

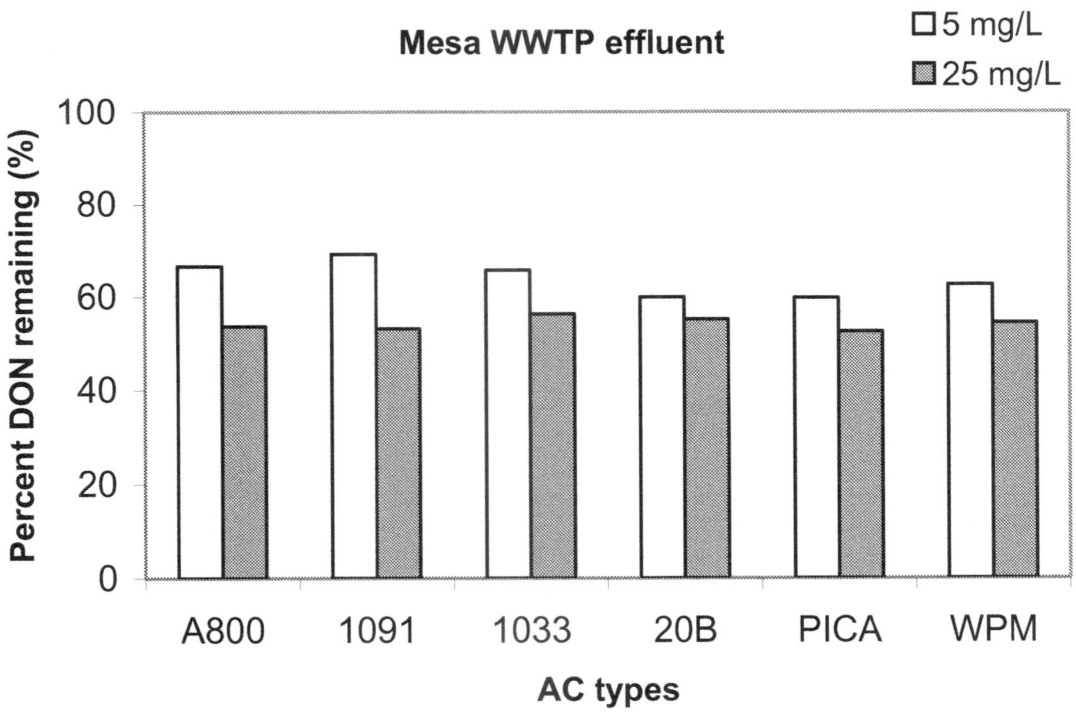

Figure 6.9 DON remaining in Mesa WWTP effluent after PAC adsorption experiments

CHAPTER 7 CHLORINATION AND CHLORAMINATION OF ORGANIC MATTER CONTAINING ORGANIC NITROGEN METHOD DEVELOPMENT

The concern for DON in drinking water was motivated by assessing its role in disinfection chemistry and disinfection by-product (DBP) formation. The formation of organic chloramine was evaluated for two reasons. First, organic chloramine can interfere with DPD methods used by utilities practicing chloramination. Since organic chloramine have a lower biocide potency than inorganic chloramine, their presence could result in underestimation of the actual disinfection credit provided under chloramination practices. Second, organic chloramine may be intermediates for carcinogenic DBP formation. While extensive DBP studies were beyond the scope of this project, a series of experiments were conducted to answer a few key questions. First, does HAN or NDMA formation depend upon the DON content of organic matter (i.e., DOC/DON ratio)? Second, can DON in organic matter serve as precursor material for NDMA and if so, which NOM fractions are yield the highest levels of NDMA?

ORGANIC CHLORAMINE FORMATION

The first set of organic chloramine (R_2NCl) experiments were conducted with free amino acids exposed to free chlorine. The free chlorine dose was 0.28 $mgCl_2$/mgDON (pH 8.2). Results for six amino acids are presented in Figure 7.1. Glycine formed organic chloramine rapidly, and they were stable for approximately 7 hours. In contrast other amino acids also rapidly produced similar levels of organic chloramine (valanine, glutamic acid, serine, threonine), but the organic chloramine rapidly decayed within 2 to 3 hours. Aspartic acid yielded lower levels of organic chloramine than glycine, and partially degraded over time. Results from aromatic amino acids (e.g., tryptophane) indicated high chlorine consumption but had low yields of organic chloramine (not shown).

After the kinetic experiments, variable free chlorine dosages (1 to 15 $mgCl_2$/mgDON) were applied to glycine and several NOM isolates (30 minutes contact time; pH 7.5). Glycine exhibited a breakpoint chlorination trend nearly identical to that of free ammonia (Figure 7.2). In contrast, chlorination of all the NOM isolates did not exhibit a breakpoint chlorination pattern. Instead free chlorine addition to NOM isolates resulted in continuously higher yields of organic chloramine. Presumably, the chlorine is oxidizing NOM and either producing organic chloramine later in the reaction process, or decreasing allowing the macromolecular organic chloramine to be detected by the methods described in Chapter 2.

NOM more enriched in organic nitrogen (e.g., lower DOC/DON ratio) produced higher levels of organic chloramine, when normalized to the amount of free chlorine consumed (Figure 7.3). Again, for the two isolates depicted in Figure 7.3, increasing free chlorine dosages (i.e., higher free chlorine consumed) resulted in higher organic chloramine yields. Varying the free chlorine to DON dose results in different rates of organic chloramine formation and decay over time (Figure 7.4 plotted on a log scale). Organic chloramine appear relatively stable for 2 to 3 hours, but decay over the time scale of days.

Free chlorine addition to NOM isolates produced organic chloramine rapidly (Figure 7.5). Over time the organic chloramine decayed. Free chlorine addition leads to more rapid organic chloramine formation than exposure to an equal concentration of preformed monochloramine (Figure 7.6). However, after approximately 1 day the organic chloramine produced from monochloramine are higher than those that would form (and decay) from free chlorine addition.

Organic chloramine formation after 10 and 100 minutes of contact time between free chlorine or preformed monochloramine and raw waters (dialysized to remove ammonia) were studied. Free chlorine and preformed monochloramine dosages were 0.3 to 0.5 mg Cl_2/mgDOC (pH 7.5). Within 100 minutes no detectable organic chloramine formation was observed upon monochloramine formation. However, the contact time may not have been long enough, given the trend observed for NOM isolates (i.e., Figure 7.6). However, organic chloramine did form rapidly upon free chlorine addition (Figure 7.7). The organic chloramine yields averaged approximately 1 mgR_2NCl as Cl_2 per mg DON present.

A series of reactions were hypothesized to explain the observed patterns in organic chloramine (R_2NCl) formation. These are listed in Table 7.1. Because organic chloramine formation appears to proceed over several hours (Figure 7.4), two types of reactive precursor sites may exist. Fast reacting sites (S_{fast}) and slower reacting sites (S_{slow}). This same type of fast and slow reacting sites have previously been proposed for chlorine demand and DBP formation (Westerhoff et al. 2001a; Westerhoff and Mash 2002), and appears consistent for organic chloramine formation. Site concentrations are probably on the order of 1 to 30 μM based upon preliminary modeling attempts. Corresponding with the reactive sites or initial oxidant (free chlorine or monochloramine) would be different rate constant values. Here only hypochlorous acid was assume to form organic chloramine, but it is plausible that free available chlorine (FAC) could also be valid. Upon formation, two mechanisms to describe the decay of organic chloramine (Figure 7.4) were proposed. Hydrolysis is probably the dominant mechanism for organic chloramine decay. However, it is possible that some oxidation of organic chloramine to other products (e.g., haloacetonitriles, trihalomethanes) may occur. In fact, the common formation pattern for haloacetonitriles approximately mimics that of organic chloramine formation and decay. As HANs decay, THMs can be produced. Ongoing research is investigating the connection between organic chloramine kinetics and HAN kinetics. Rate constants have not yet been determined for the mechanisms proposed in Table 7.1, but will be further explored as future work by the investigators on related projects.

DBP FORMATION

NOM isolates (from Table 2.3) were exposed to free chlorine or monochloramine in parallel experiments. Only chlorinated analogs formed since bromide was absent. First, the results from free chlorine addition. The haloacetonitrile yield from DOC (HAN/DOC) was relatively constant at 0.5 μg/mg, except for three colloidal NOM isolates which produced roughly 2 μg/mg (Figure 7.8). The THM yield from DOC (THM/DOC) was loosely correlated to DOC/DON ratios (Figure 7.9), with higher THM yields for NOM isolates with less organic nitrogen (i.e., higher DOC/DON ratio). The THM/HAN ratio correlated very well with the DOC/DON ratio for the NOM isolates (Figure 7.10). Free chlorine addition did not produce any detectable NDMA from any NOM isolate.

Second, addition of preformed monochloramine produced lower THM concentrations (i.e., THM/DOC yields) than free chlorine addition (Figure 7.11), but higher HAN yields (Figure 7.12). The correlation observed between THM/HAN and DOC/DON for free chlorine was less statistically significant for monochloramine experiments (Figure 7.13). NDMA formed upon monochloramine addition to NOM isolates (Figure 7.14), and appeared to form higher NDMA levels for more nitrogen enriched NOM isolates (i.e., lower DOC/DON ratios) (Figure 7.14). The colloidal NOM isolates exhibited the highest NDMA yields.

SUMMARY

Although limited experimentation was conducted on DBP formation, it focused at contrasting HAN and NDMA formation during chlorination and chloramination of 18 NOM isolates (no bromide present) with a range of DOC/DON ratios. THMs were also measured. Only chlorinated analogs formed. The following observations were made:

- NDMA did not form during chlorination of nitrogen containing NOM isolates.
- NDMA did form during chloramination of NOM isolates. This observation confirms that NOM can serve as precursor material for NDMA under chloramines conditions practiced by drinking water utilities. The colloidal NOM isolate, and other nitrogen enriched isolates, had the highest NDMA yield compared against hydrophobic acid isolates.
- The ratio of THM/HAN increased as a function of the DOC/DON ratio. Therefore, NOM isolates with higher nitrogen content had lower THM and higher HAN formation compared against NOM isolates with less nitrogen content.

Table 7.1 Hypothesized mechanisms for organic chloramine formation

Description	Reaction	Rate Constant
Organic Chloramine Formation	$S_{Fast} + HOCl \rightarrow RNH_2Cl$	$K_{HOCl\text{-}fast}$
	$S_{Slow} + HOCl \rightarrow RNH_2Cl$	$K_{HOCl\text{-}slow}$
	$S_{Fast} + NH_2Cl \rightarrow RNH_2Cl$	$K_{NH2Cl\text{-}fast}$
	$S_{Slow} + NH_2Cl \rightarrow RNH_2Cl$	$K_{NH2Cl\text{-}slow}$
Oxidation of organic chloramine	$RNH_2Cl + HOCl \rightarrow$ products	$K_{Ox,\ HOCl}$
	$RNH_2Cl + NH_2Cl \rightarrow$ products	$K_{Ox,\ NH2Cl}$
Hydrolysis of organic chloramine	$RNH_2Cl \rightarrow$ products (in H_2O)	K_H

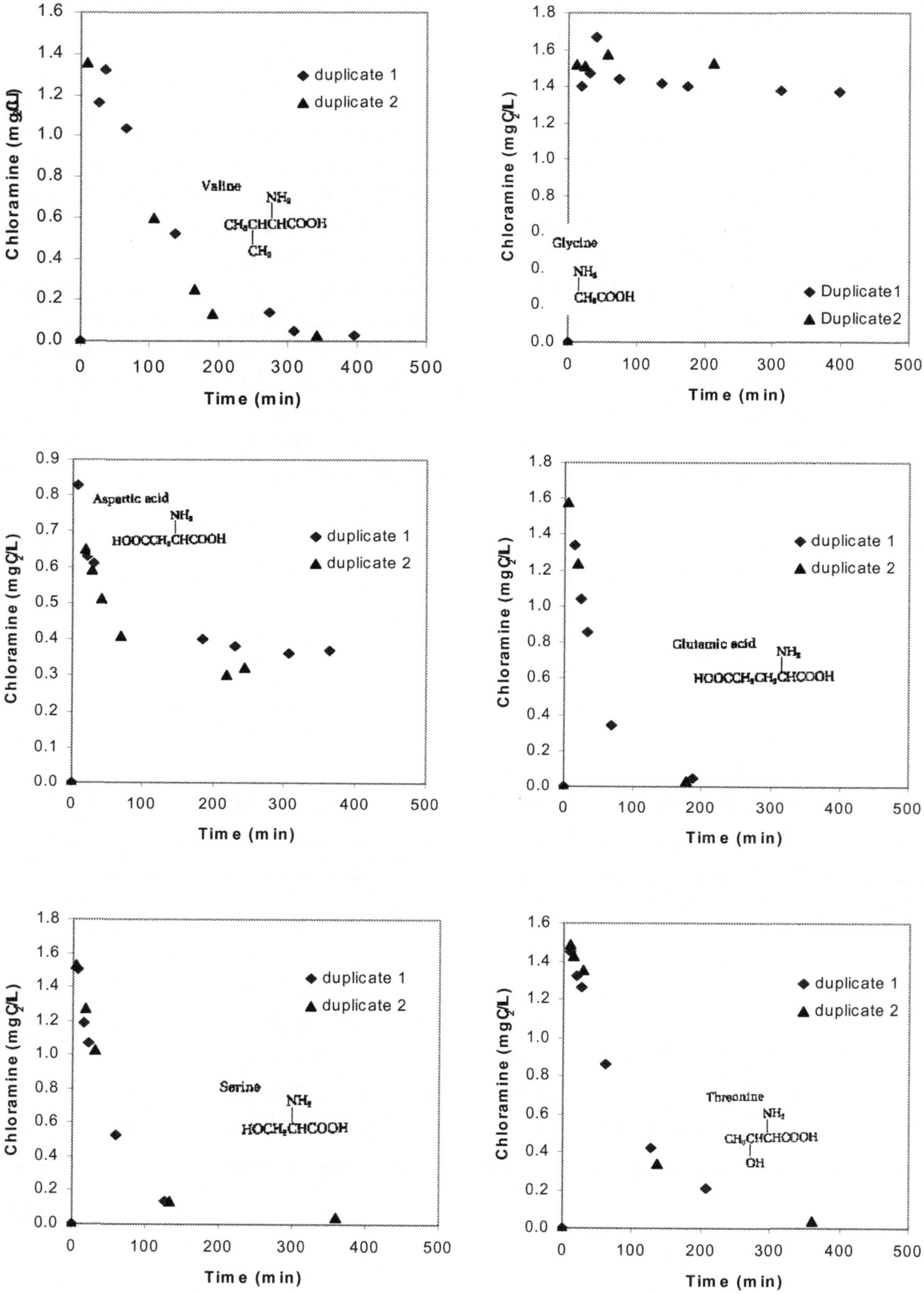

Figure 7.1 Stability of amino acids induced organic chloramine.

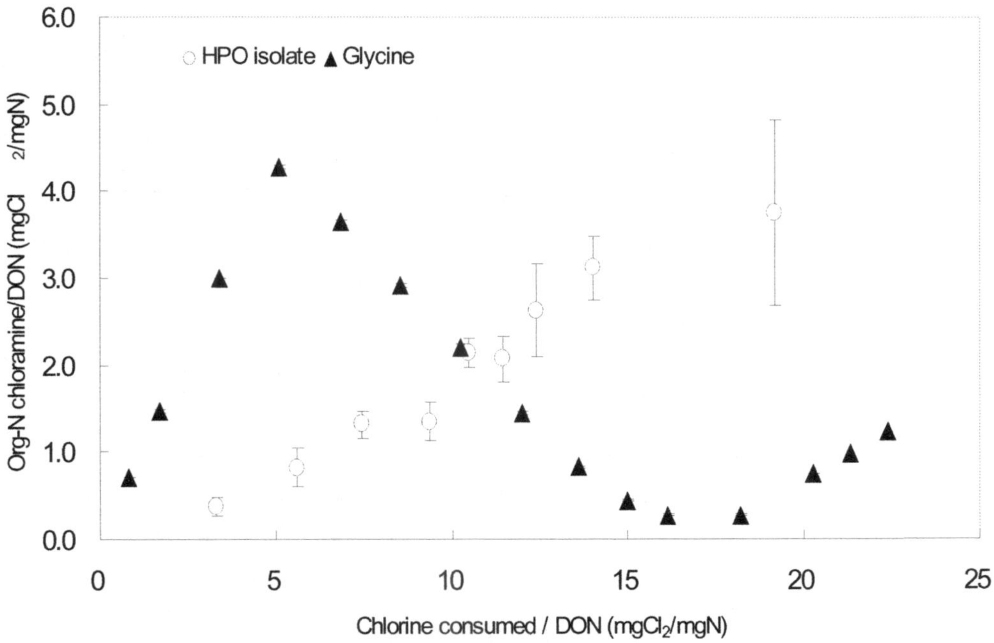

Figure 7.2 Organic chloramine formation 1 hour after chlorination of NOM isolates and glycine

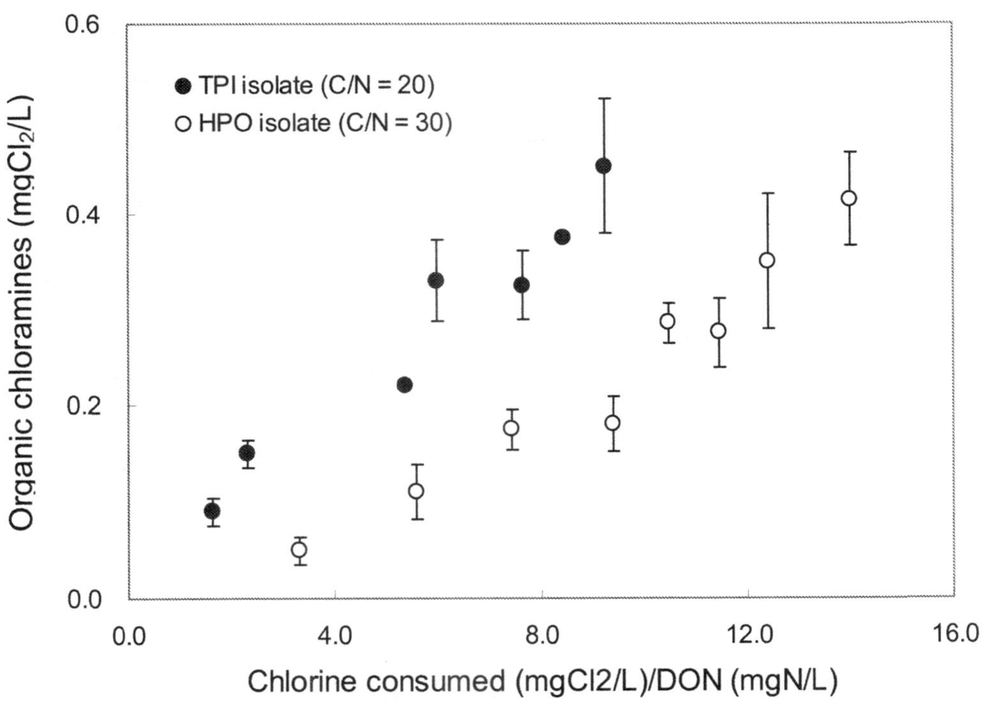

Figure 7.3 Organic chloramine formation upon chlorination of NOM isolates

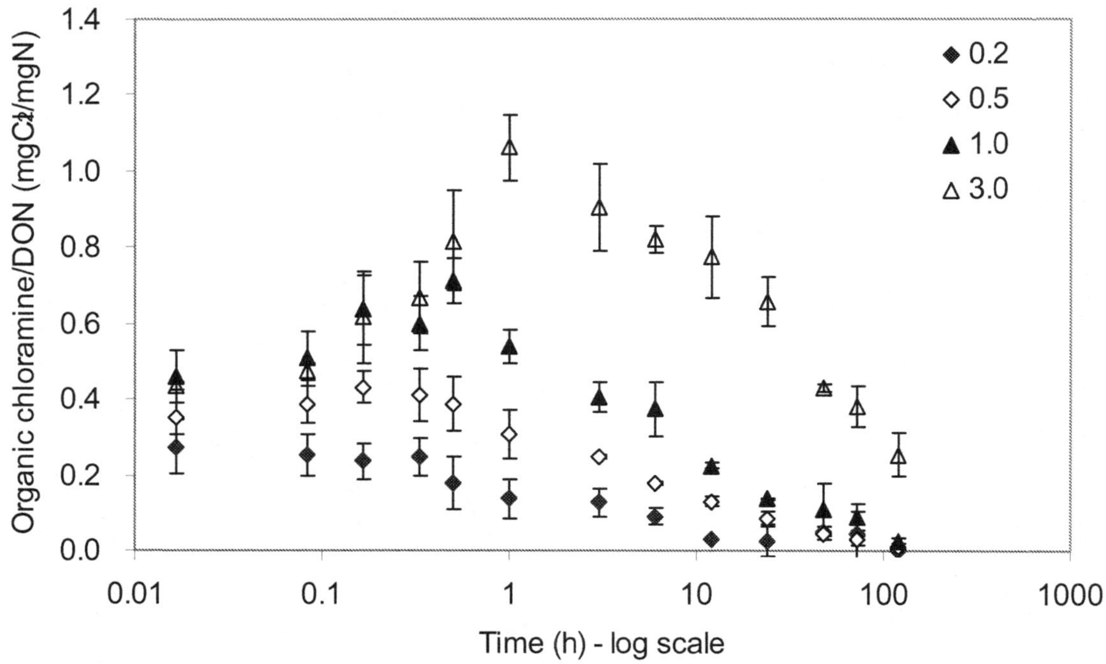

Figure 7.4 Formation and stability of organic chloramine upon chlorination of a TPI isolate with various chlorine to DON ratios [molCl$_2$/molN]

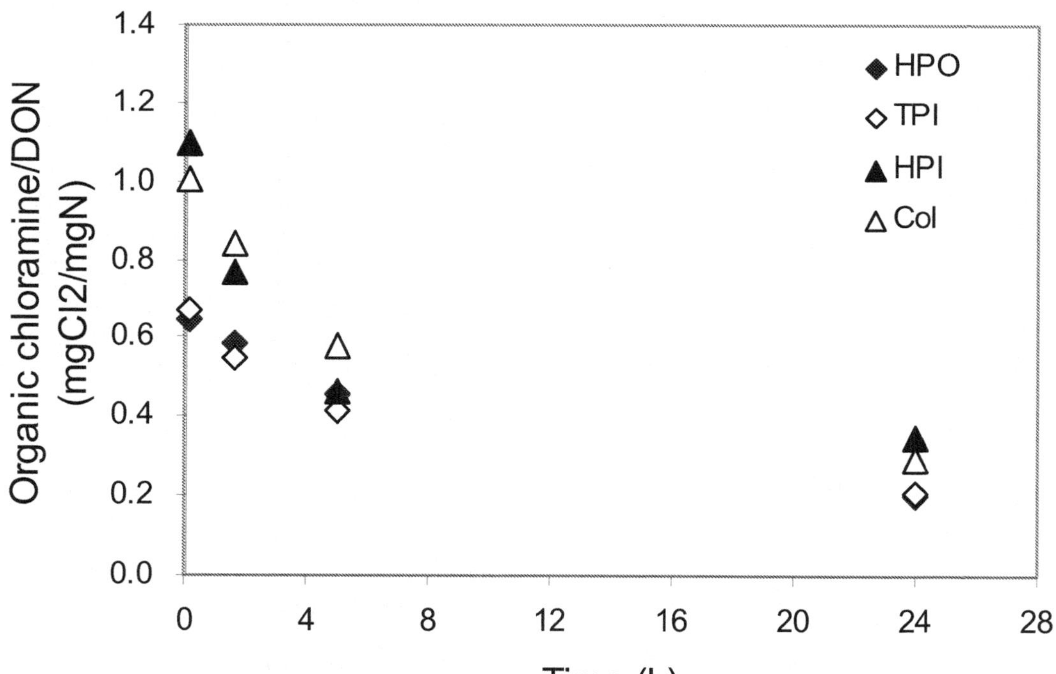

Figure 7.5 Organic chloramine formation and decay upon free chlorine addition to NOM isolates

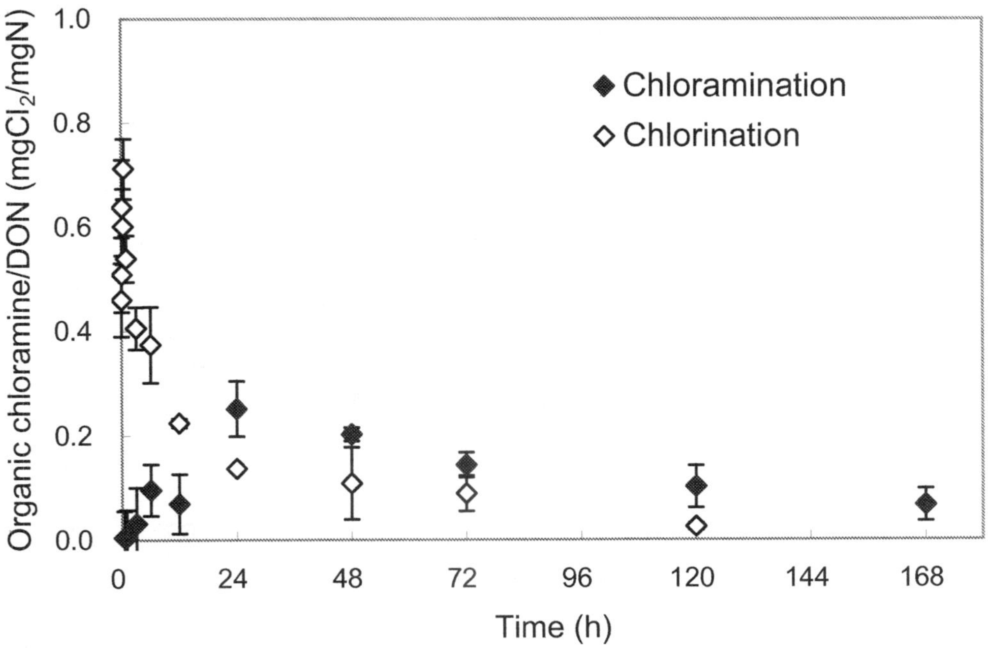

Figure 7.6 Stability of organic chloramines formed from a HPO isolate

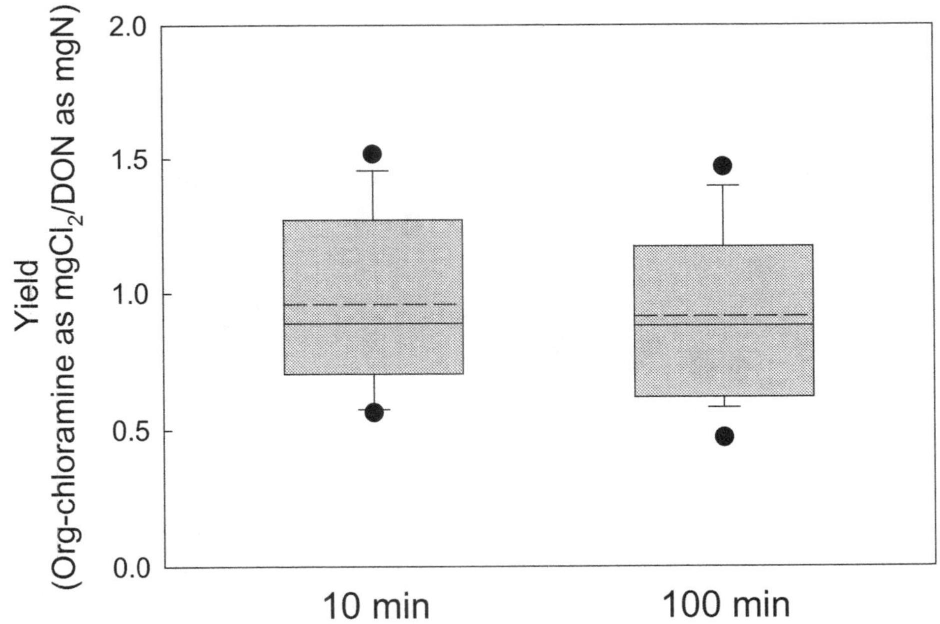

Figure 7.7 Organic chloramine formation in 10 and 100 min after chlorination of raw waters from 16 WTP; plots mean [dash line], median [solid line], 10th , 25th , 75th and 90th percentiles as vertical boxes with error bars, and symbols [•] for the 5th and 95th percentiles.

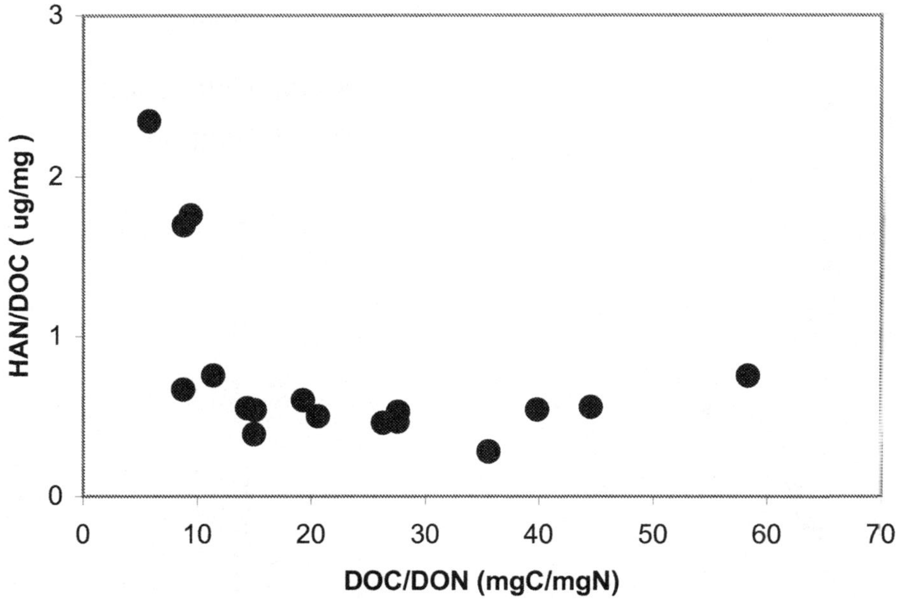

Figure 7.8 HAN formation upon chlorination of NOM isolates.

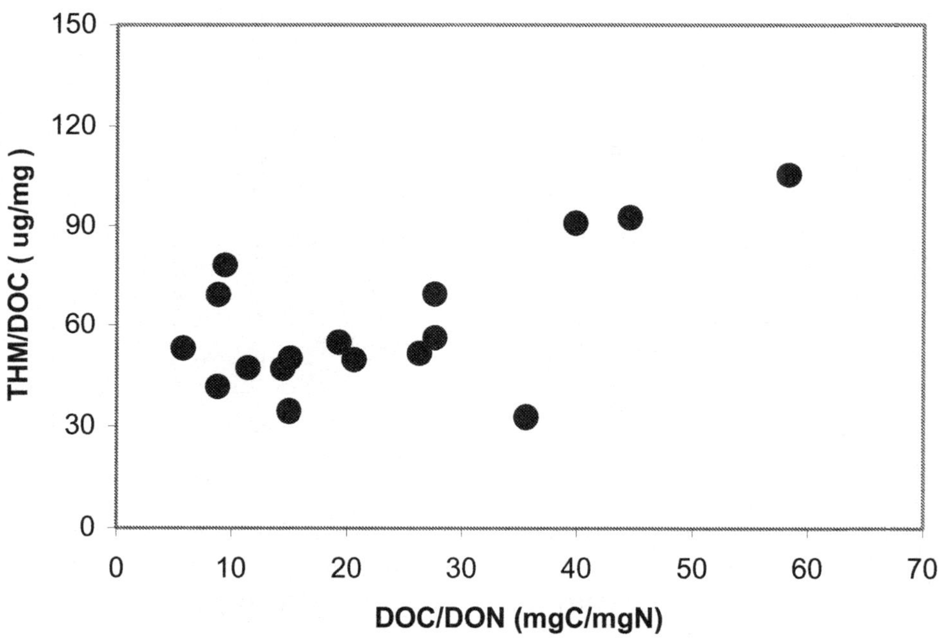

Figure 7.9 THM formation upon chlorination of NOM isolates.

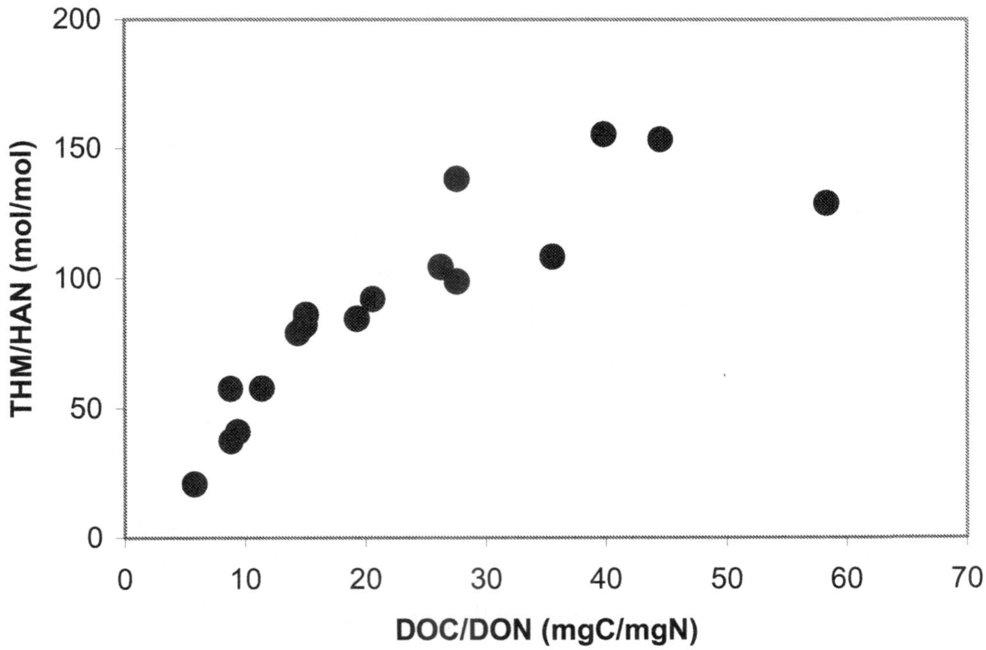

Figure 7.10 THM/HAN formation upon chlorination of NOM isolates.

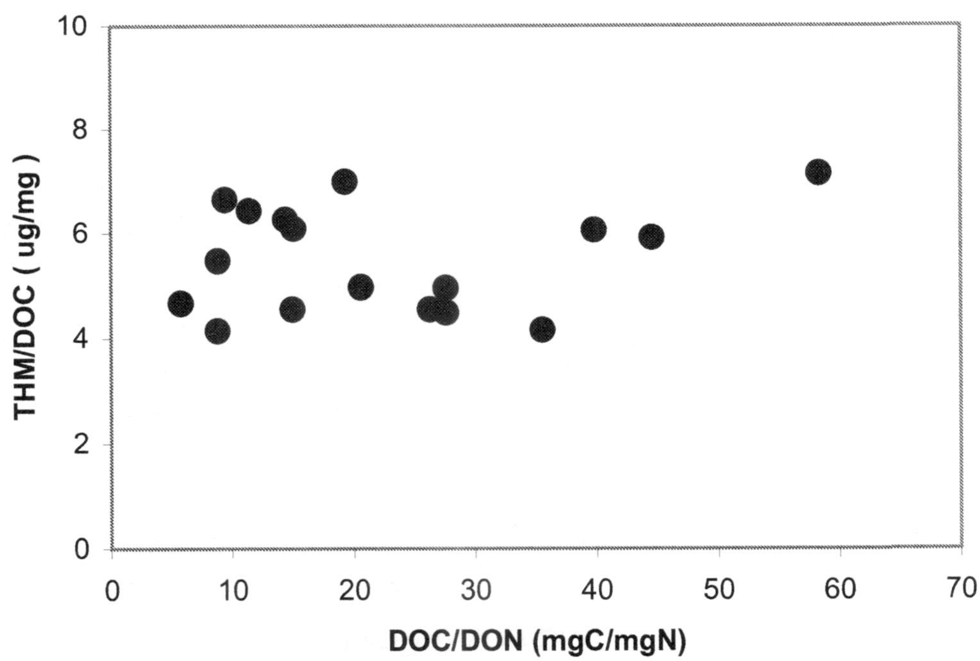

Figure 7.11 THM formation upon chloramination of NOM isolates.

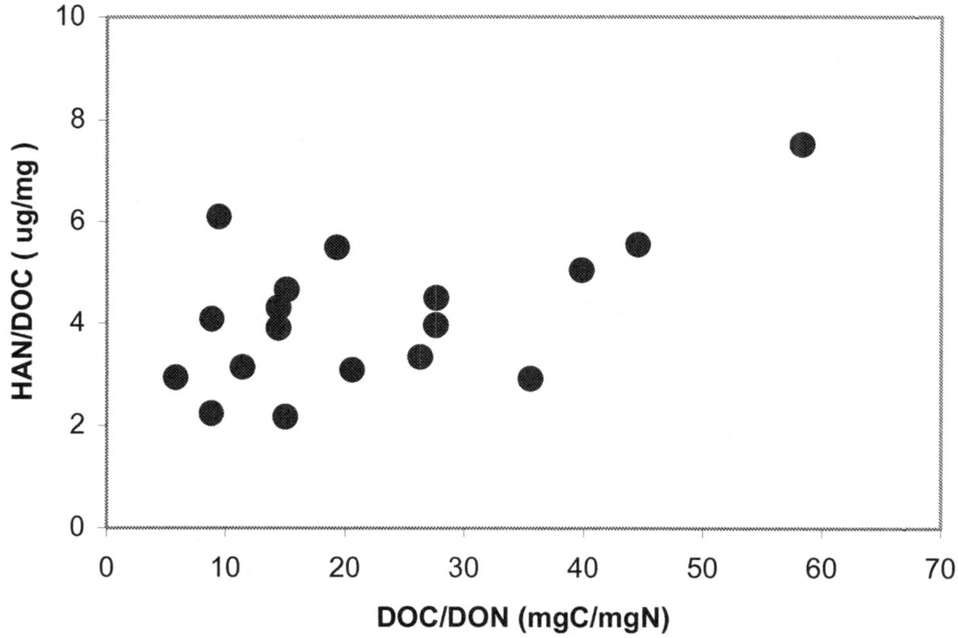

Figure 7.12 HAN formation upon chloramination of NOM isolates.

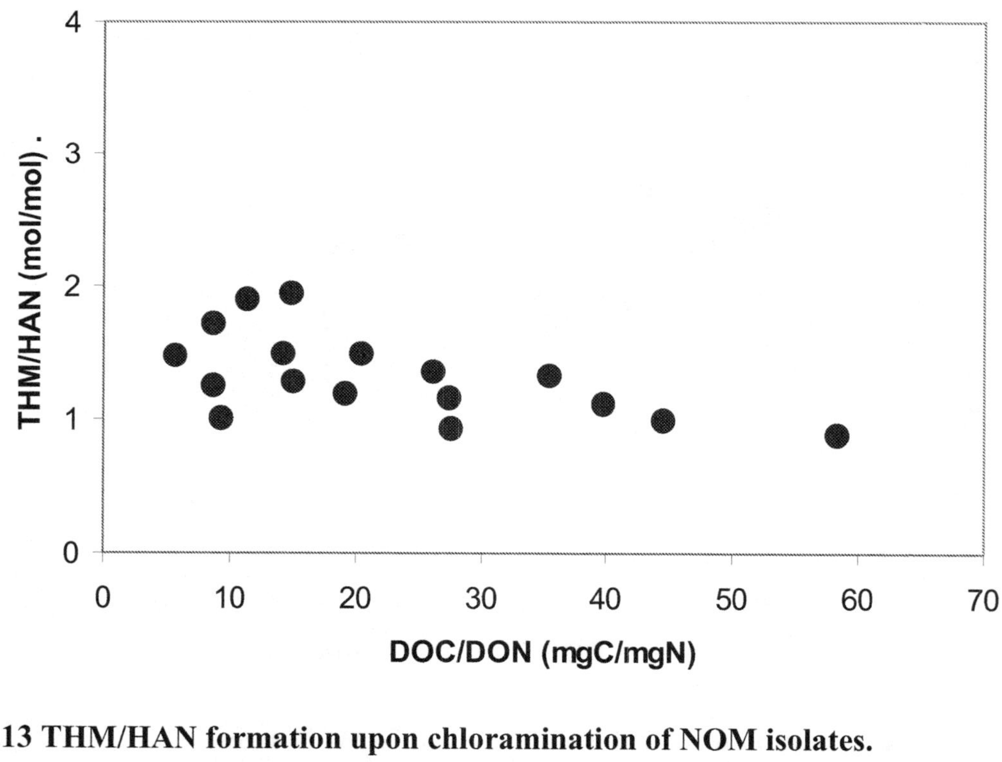

Figure 7.13 THM/HAN formation upon chloramination of NOM isolates.

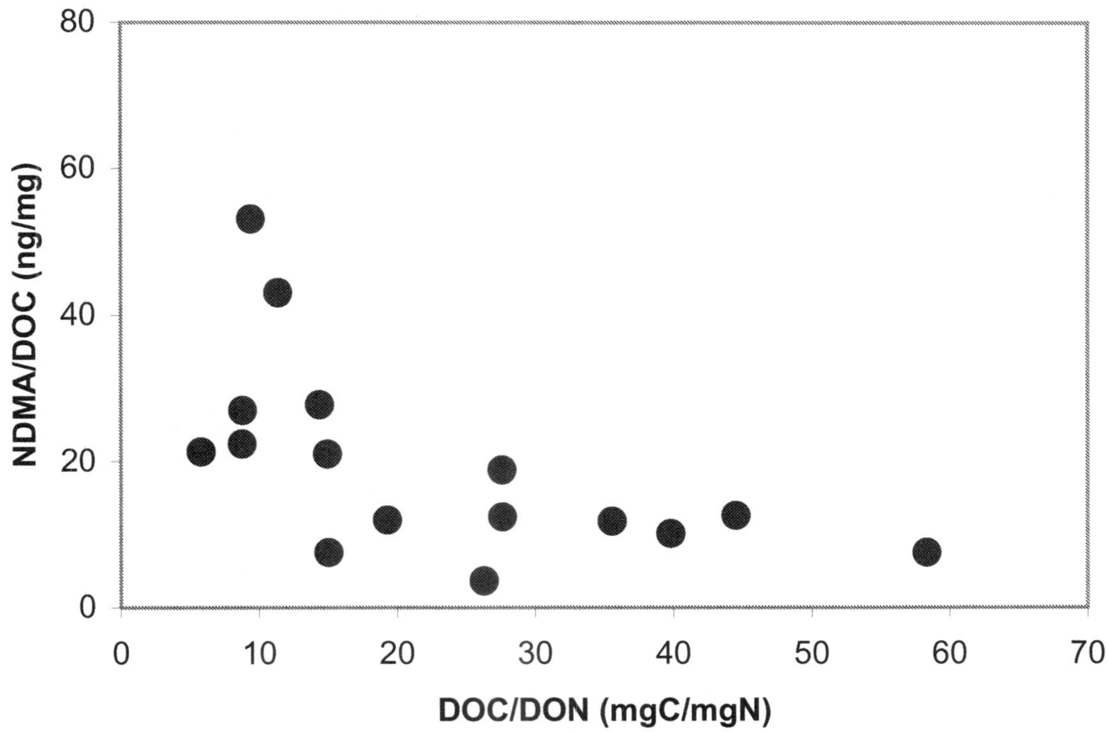

Figure 7.14 NDMA formation upon chloramination of NOM isolates

CHAPTER 8 UTILITY RECOMMENDATIONS AND FUTURE RESEARCH

This study developed a dialysis-pretreatment and validated a high-temperature oxidation analytical method that is suitable for measuring DON in drinking waters. A survey of 28 full-scale water treatment plants found an average DON concentration of 0.19 mgN/L in raw waters and 0.15 mgN/L in finished water, and DOC/DON ratios of raw water averaged 18 mgC/mgN. DON removal during water treatment closely mirrored DOC removal patterns. The molecular weight distribution of DON is similar to that for DOC, and coagulation preferentially removed the higher molecular weight DOC and DON materials. Addition of cationic polymers improved DON removal at low to moderate alum dosages. Complex (hydrolysable) amino acids and aminosugars can account for the majority of structural groups that contain organic nitrogen. Polar and colloidal NOM fractions are the most nitrogen enriched. Free chlorine and monochloramine react with nitrogenous NOM and form organic chloramine and DBPs. Organic chloramine may interfere with disinfectant measurements and serve as DBP precursors. A source for nitrogen atom in nitrogenous DBPs can be DON present in NOM. Specific findings from the study are highlighted in the Executive Summary.

RECOMMENDATIONS TO UTILITIES

This study allows utilities to measure DON and understand its role during treatment. The research will allow utilities to understand immediate and longer-lasting issues, including the following:

- **DON measurement**. In most cases TKN analysis will not provide low enough detection limits for monitoring the change in DON concentration in raw or finished waters. Utilities should monitor DON seasonally to begin collecting data that might influence disinfection practices (e.g., conversion to chloramine) or formation of emerging nitrogenous DBPs (haloacetonitriles, nitromethanes, nitrosamines).
- **Expected DON concentrations**. The average raw water DON concentration observed in this study was 0.19 mgN/L and was close to the average reported by the Investigators after analysis of USGS surface water data. Groundwaters usually have lower concentrations. Upstream agricultural activities, wastewater discharges, or significant algae productivity will result in higher DON concentrations. The average DOC/DON ratio was 18 mgDOC/mg DON and could be used to estimate the DON concentration.
- **Surrogate measurements for DON**. In general, lower SUVA tends to correlate with NOM that is enriched with organic nitrogen compared to higher NOM with a higher SUVA. Fluorescence spectroscopy appears to be a reasonable surrogate for DON, and reasonably priced instrumentation can be purchased. Specific excitation and emission wavelengths tend to correlate with protein-like material. Other surrogates (Lowry Method, Bradford Method, NanoOrange method) have inadequate MDLs or potential inorganic interferences.
- **Composition of DON**. Large macromolecular compounds contain amino acid and aminosugar components, which account for up to 60% of the DON. The concentration of free amino acids are less than 5% of the DON. NOM fractions have different nitrogen content. The colloidal NOM fraction generally has the highest nitrogen content, and is

indicative of biological activity. More polar (less hydrophobic) fractions are also nitrogen enriched, so humic substances tend to have less DON than NOM not retained on XAD8 resins.

- **Removal of DON**. Coagulation with metal salts have equal or slightly lower DON percentage removals compared to DOC removals. High molecular weight DON, and DOC, is preferentially removed during coagulation. To improve DON removal at low alum dosages, cationic polymers can be used. DON of biological origin (e.g., wastewater discharges) tends to be difficult to coagulate, but can exhibit a moderate tendency to adsorb to activated carbon (e.g., GAC). Membrane tests were not specifically conducted during this study, but based upon the ultrafiltration molecular weight fraction conducted, DON and DOC distributions among molecular weight fractions were quite similar. Therefore, membrane processes capable of removing DOC should also remove DON.
- **Chlorination or Chloramination of DON**. Organic chloramine form due to reactions between DON and these oxidants. Organic chloramine interfere with Total DPD colorimetric measurements of combined chlorine residuals. The indophenol method (Hach MonoChlorF reagent) did not exhibit interferences from organic chloramine, and should be used for monitoring. Utilities may want to pursue states to adopt the indophenol method for monitoring monochloramine. Organic chloramine formation in the presence of free chlorine occurs rapidly within a few minutes, whereas formation in the presence of preformed monochloramine is much slower and did not reach a maximum concentration until nearly 24 hours after exposure to the monochloramine. This is significant, because utilities using short free chlorine contact times for in-plant microbial inactivation will form organic chloramine, before ammonia addition to form monochloramine as a post disinfectant. Once formed organic chloramine degrade slowly over days due to hydrolysis.
- **Organic Chloramine.** What is the level of organic chloramine in water distribution systems, and do they lead to negative water quality impacts? For example, the Investigators can hypothesize a sequence of events leading to nitrification in water distributions for utilities practicing chloramination. An algae bloom occurs in late summer in a watershed. The DOC/DON ratio decreases, but the absolute DOC concentration may only change slightly. DON concentrations increase in treated water. Free chlorine is added for disinfection, forming organic chloramine. Then free ammonia is added to form inorganic chloramine for post disinfection, but organic chloramine are already present and continue to form slowly due to reactions between DON and monochloramine. With travel time away from the WTP total DPD measurements indicate a 1 $mgCl_2/L$ combined residual, but > 75% of that residual could be due to organic chloramine or dichloramine. In the presence of these weak biocides, nitrifying bacteria proliferate. The remaining monochloramine reacts with the biofilm (nitrogen rich) and may form additional organic chloramine that remain detectable by total DPD detection. Then nitrification occurs and all residual is lost. The common remedy then includes several weeks of free chlorination. Does organic chloramine form during this period as free chlorine reacts with the biofilms? Does NDMA or other nitrogenous form? Much of this potential problem could have been avoided if utilities utilize and state regulators adopt the indophenol method for monochloramine detection and monitoring in water distribution systems.

- **Nitrogenous DBPs**. Free chlorination of NOM formed HANs and THMs, whereas chloramination formed higher HAN but lower THM concentrations. In the presence of free chlorine, HANs can serve as intermediates and ultimately form THMs. Free chlorination of NOM isolates did *not* form NDMA, but NDMA did form in the presence of monochloramine. Overall DON is a precursor for nitrogenous DBPs (HAN, NDMA) as well as currently regulated DBPs (THMs). Through monitoring DON, a better appreciation for the presence of nitrogenous DBP precursors will be gained.

RECOMMENDATIONS FOR FUTURE RESEARCH

The Investigators assisted in organizing sessions and presenting results at the 2004 AWWA Annual Conference and Water Quality and Technology Conferences. This project has focused AwwaRF and researchers on the significance of nitrogenous DBPs and DBP precursors. As research in this area progresses, some important questions that remain unanswered include the following:

- Has inadequate hydrolysis techniques been the limiting factor in accounting for composition of DON in surface waters? If so, then amino acids and aminosugars account for the majority of the DON. However, methods to quantify diverse heterocyclic nitrogen compounds remain undeveloped, and are limited to quantification of specific compounds (pesticides, chlorophyll, algae toxins, etc).
- Can an easy-to-apply method be developed to estimate the colloidal organic matter fraction in water? Colloidal NOM is nitrogen enriched and reacts to form organic chloramine, HANs, THMs, and NDMA. However, unlike XAD chromatography a widely applied operational definition of colloidal matter has not been adopted. Colloidal NOM is thought to consist of cell wall-like fragments and as such may be amenable to GC-MS or other analysis currently applied for endotoxin analysis.
- While dialysis pretreatment allows for accurate DON determination in most WTP raw and finished waters and reclaimed wastewaters, the determination still relies upon measurement of inorganic species and total dissolved nitrogen measurements. Can a direct DON method be developed? The current dialysis pretreatment method may still be inadequate for waters with very high nitrate or ammonia concentrations (e.g., wastewater with primary treatment or partial nitrification or denitrification). While several techniques are reported in this report, and others evaluated in the Investigators laboratories, a universal direct DON method still is not available.
- Could the dialysis pretreatment method serve as a means for detailed studies of DBP formation? Dialysis allows passage of bromide and other chlorine reactive inorganic compounds while causing minimal alteration or loss of NOM, when compared against harsh chemical isolation techniques for example. Therefore, the role of bromide on DBP formation could be explored using dialysis NOM in the absence of bromide.
- What is the role of organic chloramine in DBP and total organic halide (TOX) formation? In this report the yields and formation/degradation kinetics of organic chloramine are presented. However, direct linkage to DBP formation was beyond the scope of work. In addition, the Investigators have wondered if organic chloramine would be recovered and detected during TOX analysis. If so, then a water with 0.2 mgN/L of DON could form 0.1 to 0.5 mgCl_2/L of organic chloramine when exposed to free chlorine or

monochloramine. If recovered as TOX, this could account for several hundred micrograms per liter as chloride.

- Does bromide form organic chloramine? If so, how can they be quantified and are they stable in water? Currently there are limited direct analytical measurements available for quantifying organic chloramine, other than free amino-acid based organic chloramine by HPLC. It is likely that most organic chloramine form during reaction with macromolecular structures containing amino acids. Which of these hydrolysable amino acids are most reactive with free chlorine, free bromine, or monochloramine? Does dichloramine react and form organic chloramine?
- What is the level of organic chloramine in water distribution systems, and do they lead to negative water quality impacts? Do organic chloramine form during this period as free chlorine reacts with the biofilms? Does NDMA or other nitrogenous form? Much of this potential problem could have been avoided if utilities and state regulators adopt the indophenol method for monochloramine detection and monitoring in water distribution systems.

REFERENCES

Aiken, G., and Cotsaris, E. (1995). "Soil and Hydrology - Their Effect On NOM." *Journal American Water Works Association*, 87(1), 36-45.

Ammann, A. A., Ruttimann, T. B., and Burgi, F. (2000). "Simultaneous determination of TOC and TNb in surface and wastewater by optimized high temperature catalytic combustion." *Water Research*, 34(14), 3573-3579.

APHA, AWWA, and WEF. (1998). *Standard Methods for the Examination of Water And Wastewater*, American Public Health Association, Washington, DC.

AwwaRF. (1993). "Natural Organic Matter Workshop (Chamonix, France)." AwwaRF, Denver, CO.

AwwaRF. (2001). "Microbial / Disinfection By-Product Workshop Final Report." , American Water Works Association Research Foundation, Denver, CO.

Chen, W., Westerhoff, P., Leenheer, J. A., and Booksh, K. (2003). "Fluorescence excitation-emission matrix regional integration to quantify spectra for dissolved organic matter." *Environmental Science and Technology*, 37, 5701-5710.

Coble, P. G. (1996). "Characterization of marine and terrestrial DOM in seawater using excitation emission matrix spectroscopy." *Marine Chemistry*, 51(4), 325-346.

Croue, J.-P., Korshin, G. V., and Benjamin, M. (1999). "Characterization of Natural Organic Matter in Drinking Water." AwwaRF, Denver, CO.

Dennehy, K. F., Litke, D. W., Tate, C. M., Qi, S. L., McMahon, P. B., Bruce, B. W., Kimbrough, R. A., and Heiny, J. S. (1998). "Water Quality in the South Platte River Basin, Colorado, Nebraska, and Wyoming, 1992-95." *1167*, U.S. Geological Survey, Denver, CO.

Devore, J. L. (1995). *Probability and statistics for engineering and the sciences*, Brooks/Cole Publishing, Pacific Grove, CA.

Esparza-Soto, M., and Westerhoff, P. K. (2001). "Fluorescence spectroscopy and molecular weight distribution of extracellular polymers from full-scale activated sludge biomass." *Water Science and Technology*, 43(6), 87-95.

Fleck, A., and Munro, H. N. (1965). "The Determination of Organic Nitrogen in Biological Materials." *Clinica Chimica Acta*, 11, 2-12.

Her, N., Amy, G., Foss, D., Cho, J., Yoon, Y., and Kosenka, P. (2002a). "Optimization of method for detecting and characterizing NOM by HPLC-size exclusion chromatography with UV and on-line DOC detection." *Environmental Science & Technology*, 36(5), 1069-1076.

Her, N., Amy, G., Foss, D., and Cho, J. W. (2002b). "Variations of molecular weight estimation by HP-size exclusion chromatography with UVA versus online DOC detection." *Environmental Science & Technology*, 36(15), 3393-3399.

Her, N., Amy, G., McKnight, D., Sohn, J., and Yoon, Y. M. (2003). "Characterization of DOM as a function of MW by fluorescence EEM and HPLC-SEC using UVA, DOC, and fluorescence detection." *Water Research*, 37(17), 4295-4303.

Her, N., Amy, G., Park, H. R., and Song, M. (2004). "Characterizing algogenic organic matter (AOM) and evaluating associated NF membrane fouling." *Water Research*, 38(6), 1427-1438.

Jones, B. M., and Daughton, C. G. (1985). "Chemi-Luminescence Vs Kjeldahl Determination of Nitrogen in Oil-Shale Retort Waters and Organonitrogen Compounds." *Analytical Chemistry*, 57(12), 2320-2325.

Jones, L. J., Haugland, R. P., and Singer, V. L. (2003). "Development and characterization of the NanoOrange protein quantitation assay: a fluorescence-based assay of proteins in solution." *Biotechniques*, 34(4), 850-861.

Jorgensen, T. C., and Weatherley, L. R. (2003). "Ammonia removal from wastewater by ion exchange in the presence of organic contaminants." *Water Research*, 37(8), 1723-1728.

Jung, J. Y., Chung, Y. C., Shin, H. S., and Son, D. H. (2004). "Enhanced ammonia nitrogen removal using consistent biological regeneration and ammonium exchange of zeolite in modified SBR process." *Water Research*, 38(2), 347-354.

Klotz, J. B., and Pyrch, L. A. (1999). "Neural tube defects and drinking water disinfection by-products." *Epidemiology*, 10(4), 383-390.

Kogel-Knabner, I. (1997). "13C and 15N NMR spectroscopy as a tool in soil organic matter studies." *Geoderma*, 80(243-270).

Krasner, S. W., Pastor, S., Chinn, R., Sclimenti, M. J., Weinberg, H. S., Richardson, S. D., and Thruston, A. D. 2001. "The occurrence of a new generation of DBPs (Beyond the ICR)." *American Water Works Association WQTC*, Nashville, TN, 24.

Lee, W., and Westerhoff, P. (Accepted). "Dissolved organic nitrogen measurement using dialysis pretreatment." *Environmental Science and Technology*.

Levy, G. C., and Lichter, R. L. (1979). *Nitrogen-15 Nuclear Magnetic Resonance Spectroscopy (Chapters 3 and 6)*, Wiley, New York.

Lowry, O. H., Rosenbrough, N. J., Farr, A. L., and Randall, R. J. (1951). "Protein measurement with the folin phenol reagent." *J. Biol. Chem.*, 193, 265-275.

MacKenzie, K. J. D., and Smith, M. E. (2002). "Multinuclear Solid-state NMR of Inorganic Materials." , Section 9.3.2.

Martens, D. A., and Loeffelmann, K. L. (2003). "Soil Amino Acid Composition Quantified by Acid Hydrolysis and Anion Chromatography–Pulsed Amperometry." *J. Agric. Food Chem.*, 51, 6521-6529.

Mitch, W. A., and Sedlak, D. L. (2004). "Characterization and fate of N-nitrosodimethylamine precursors in municipal wastewater treatment plants." *Environmental Science & Technology*, 38(5), 1445-1454.

Nguyen, M. L., Baker, L. A., and Westerhoff, P. (2002). "DOC and DBP precursors in western US watersheds and reservoirs." *Journal of American Water Works Association*, 94(5), 98-112.

Nolan, B. T., and Stoner, J. D. (2000). "Nutrients in groundwaters of the conterminous United States 1992-1995." *Environmental Science & Technology*, 34(7), 1156-1165.

Parkin, G. F., and McCarty, P. L. (1981). "A comparison of the characteristics of soluble nitrogen in untreated and activated sludge treated wastewaters." *Water Research*, 15, 139-149.

Plewa, M. J., Kargalioglu, Y., Vankerk, D., Minear, R. A., and Wagner, E. D. (2000). "Development of a quantitative comparative cytotoxicity and genotoxicity assays for environmental hazardous chemicals." *Water Science and Technology*, 42(7-8), 109-116.

Plewa, M. J., Kargalioglu, Y., Vankerk, D., Minear, R. A., and Wagner, E. D. (2002). "Mammalian cell cytotoxicity and genotoxicity analysis of drinking water disinfection by-products." *ENVIRONMENTAL AND MOLECULAR MUTAGENESIS*, 40(2), 134-142.

Reckhow, D. A., Singer, P. C., and Malcolm, R. L. (1990). "Chlorination of Humic Materials - By-Product Formation and Chemical Interpretations." *Environmental Science & Technology*, 24(11), 1655-1664.

Richardson, S. D., Thruston, A. D., Caughran, T. V., Chen, P. H., Collette, T. W., and Floyd, T. L. (1999). "Identification of new drinking water disinfection byproducts formed in the presence of bromide." *Environmental Science & Technology*, 33(19), 3378-3383.

Sharp, J. H., Rinker, K. R., Savidge, K. B., Abell, J., Benaim, J. Y., Bronk, D., Burdige, D. J., Cauwet, G., Chen, W. H., Doval, M. D., Hansell, D., Hopkinson, C., Kattner, G., Kaumeyer, N., McGlathery, K. J., Merriam, J., Morley, N., Nagel, K., Ogawa, H., Pollard, C., Pujo-Pay, M., Raimbault, P., Sambrotto, R., Seitzinger, S., Spyres, G., Tirendi, F., Walsh, T. W., and Wong, C. S. (2002). "A preliminary methods comparison for measurement of dissolved organic nitrogen in seawater." *Marine Chemistry*, 78(4), 171-184.

Smith, R. A., Alexander, R. B., and Schwarz, G. E. (2003). "Natural background concentrations of nutrients in streams and rivers of the conterminous United States." *Environmental Science & Technology*, 37(14), 3039-3047.

Taylor, J. S., Thompson, D. M., and Carswell, J. K. (1987). "Applying membrane processes to groundwater sources for trihalomethane precursor control." *Journal of American Water Works Association*, 79(8), 72-82.

Turner, G. L., Kirkpatrick, R. J., Risbud, S. H., and Oldfield, E. (1987). *Amer. Ceram. Soc. Bull.*, 70, 675.

USEPA. (1999). *Code of Federal Regulations, vol. 40, Part 136, Appendix B*, , Washington DC.

Westerhoff, P., Q. He, G. Amy, D. Reckhow, M. Addison, W. Swanson, Z. Chowdhury, J. McClellan. (2001a). "Comparison of chlorine and DBP predictions by mechanistic and empirical models," AWWA Annual Conference, CD-ROM Paper # ACE01-MON09-03, 10 pages, Washington, D.C.

Westerhoff, P., Chen, W., and Esparza, M. (2001b). "Fluorescence analysis of a standard fulvic acid and tertiary treated wastewater." *Journal of Environmental Quality*, 30(6), 2037-2046.

Westerhoff, P. and H. Mash. (2002). "Dissolved organic nitrogen in drinking water supplies: A review," *Journal of Water Supply Research and Technology-Aqua*, 51:8: 415-448.

Westerhoff, P., Reckhow, D., Amy, G., Chowdhury, Z. (2002). "Mechanistic-based Disinfectant and Disinfectant By-Product Models," USEPA Grant #R826831-01-0, 46 pages, Final Report.

ABBREVIATIONS

AA	Amino acid
AOC	Assimable organic carbon
Arg	Arginine
Ala	Alanine
ASP	Aspartic acid
Awwa	American Water Works Association
AwwaRF	Awwa Research Foundation
AQC	6-aminoquinolyl-N-hydroxysuccinimidyl carbamate
AS	Aminosugar
AZ	Arizona
BSA	Bovine serum albumin
CAP	Central Arizona Project
CE	Cellulose ester
Cl_2	Chlorine
Cys	Cystine
^{o}C	Degree Celsius
Da	Dalton
DBP	Disinfection byproduct
DIN	Dissolved inorganic nitrogen
DOC	Dissolved organic carbon
DOM	Dissolved organic matter
DON	Dissolved organic nitrogen
EEM	Excitation-emission matrix fluorescence
EfOM	Effluent organic matter
FA	Fulvic acids
GAC	Granular activated carbon
GC/MS	Gas chromatography mass-spectrometry
GF/F	Glass fiber filter (fine)
Glu	Glutamic acid
Gly	Glycine
HAA	Haloacetic acid
HAN	Haloacetonitrile
HCl	Hydrochloric acid
His	Histidine
HPIA+N	Hydrophilic acids plus neutrals
HPLC	High performance liquid chromatography
HPSEC	High performance size exclusion chromatography

HPO	Hydrophobic organic matter
HPO-A	Hydrophobic acids
HTO	High temperature oxidation
IHSS	International humic substances society
IL	Illinois
Ile	Isoleucine
IN	Indiana
Jour.	Journal
KY	Kentucky
L	Liter
Leu	Leucine
Lys	Lysine
m	meter
MA	Massachusetts
MCL	Maximum contaminant level
MDL	Minimum detection level
Met	Methionine
MI	Michigan
min	Minute
MO	Missouri
MSA	methanesulfonic acid
MW	Molecular weight
MWCO	Molecular weight cut-off
NDN	Nitrifying-denitrifying
NH_4^+	Ammonium ion
NH_3	Ammonia
NO_2^-	Nitrite
NO_3^-	Nitrate
NO_x	Nitrous oxides
NAWQA	USGS National Water Quality Assessment
NDMA	Nitrosodimethylamine
NOM	Natural organic matter
NTU	Nephlometric Turbidity Units
NV	Nevada
OPA	Orthophtaldialdehyde
PA	Pennsylvania
PAC	Powder activated carbon
Phe	Phenylalanine

PON	Particulate organic nitrogen
Pro	Proline
R_2NCl	Organic chloramine
RSD	Residual standard deviation
Ser	Serine
SMP	Soluble microbial product
SUVA	Specific ultraviolet absorbance at 254 nm
TDS	Total dissolved solids
TF	Trickling filter
THM	Trihalomethane
Thr	Threonine
TKN	Total Kjeldahl nitrogen
TOC	Total organic carbon
TPI	Transphilic organic matter
TPI-A	Transphilic acids
TPI-N	Transphilic neutrals
TX	Texas
Tyr	Tyrosine
URI	Ultraviolet ratio index
USA	United States of America
USGS	United States Geologic Survey
UV	Ultraviolet absorption
UVA_{254}	Ultraviolet absorption at 254 nm
VA	Virginia
Van	Valine
WCO	Wet chemical oxidation
WI	Wisconsin
WTP	Water treatment plant
WWTP	Wastewater treatment plant

IWA Publishing's authorised EU representative for General Product Safety Regulations is Diane D'Arras, 15 rue Duret, 75116 Paris, France, e-mail: safety@iwap.co.uk.

Printed and bound by CPI Group (UK) Ltd, Croydon, CR0 4YY
17/09/2026
02218799-0019